The Open University

A Second Level Course

ELECTROMAGNETICS AND ELECTRONICS

Signal Processing I
Transient Response   Unit 6

Signal Processing II
Sinusoidal Response   Unit 7

Prepared by a Course Team from the Faculties of Technology and Science

THE OPEN UNIVERSITY PRESS

The Electromagnetics and Electronics Course Team

J. J. Sparkes (Chairman)

G. R. Alexander	(Electronics)
G. Bellis	(Equipment)
E. Braley-Smith	(Editorial)
P. F. Chapman	(Physics)
P. R. Cooper	(Course Assistant)
W. A. Cooper	(Physics)
G. P. Copp	(Editorial)
D. I. Crecraft	(Electronics)
J. G. Gregory	(Editorial)
A. B. Jolly	(BBC)
R. Loxton	(Electronics)
G. D. Moss	(Educational Technology)
G. Smol	(Electronics)
L. A. Suss	(Administration)
M. Weatherley	(BBC)

The following people acted as consultants for certain components of the course.
E. A. Faulkner and members of the J. J. Thomson Laboratory, Reading
R. Knight

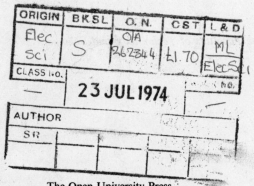

The Open University Press
Walton Hall, Bletchley, Buckinghamshire

First published 1972

Copyright © 1972 The Open University

All rights reserved. No part of this work may be reproduced in any form, by mimeograph or any other means, without permission in writing from the publishers.

Designed by The Media Development Group of the Open University.

Printed in Great Britain by
Martin Cadbury Printing Group

SBN 335 04003 9

This text is one in a series of units that make up the correspondence element of an Open University Second Level course. The complete list of units in the course is given at the end of this text.

For general availability of supporting material referred to in this book, please write to the Director of Marketing, The Open University, Walton Hall, Bletchley, Buckinghamshire.

Further information on Open University courses may be obtained from The Admissions Office, The Open University, P.O. Box 48, Bletchley, Buckinghamshire.

1.1

Unit 6

Signal Processing I. Transient Response

CONTENTS

	Aims and objectives	6
6.1	Introduction	7
6.1.1	Some network conventions and notation	8
6.2	Kirchhoff's laws	10
6.2.1	Kirchhoff's current law	10
6.2.2	An extension of Kirchhoff's current law	11
6.2.3	Kirchhoff's voltage law	12
6.3	Resistive circuits	14
6.3.1	Summary of sections 6.2 and 6.3	20
6.4	Resistance–capacitance circuits	21
6.5	Resistance–inductance circuits	24
6.6	Other single time constant circuits	27
6.6.1	An unobvious example. A transformer with resistive load	31
6.6.2	Summary of sections 6.4, 6.5 and 6.6	33
6.7	A nonlinear circuit. The dc restorer	34
6.8	*RLC* circuits	37
6.8.1	Summary of section 6.8	41
6.9	Circuit transient or natural response	43
6.10	Summary of Unit 6	45
	Self-assessment questions	47
	Self-assessment answers and comments	48

Aims

1 To introduce the analysis of electrical components connected together in simple circuits.

2 To convey an understanding, both mathematically and intuitively, of how the simplest circuits (R, RC, RL, RLC) process some simple signals (step functions, square waves).

3 To apply this understanding to such applications as diode clamping circuits and transformers with a resistive load.

Objectives

When you have completed this unit you should be able to:

1 Define, or recognize adequate definitions of, or distinguish between true and false statements concerning each of the following terms or concepts

characteristic equation	linear and nonlinear resistive devices
clamping circuit	natural frequency
critical damping	node
current divider rule	overdamped
damped oscillation	resistive circuit
dc restoration	rise time
duality	single time constant circuit
forced response	square wave
ideal current source	steady state
ideal transformer	time constant
ideal voltage source	transient response
Kirchhoff's current law	underdamped
Kirchhoff's voltage law	voltage divider rule

2 State Kirchhoff's laws and apply them to any circuit.

3 Determine whether a given device or circuit is or is not resistive.

4 Describe the voltage and current waveforms in a simple RC or RL circuit when a step in voltage or current, respectively, is applied to them.

5 Determine whether a given circuit is a single time constant circuit. If so, write the expression for its current and voltage waveforms when a step in voltage or current, respectively, is applied, but without solving the circuit's differential equation.

6 Use the principle of duality to find the voltages and currents in a circuit, given the voltages and currents in its dual circuit.

7 Draw the waveforms resulting from the application of a square wave to an RC or RL circuit and an RC circuit with dc restoration.

8 Describe the waveforms resulting from the application of a step voltage to an RLC circuit for varying values of R.

9 Describe the waveforms resulting from the application of a step in voltage to a transformer with a resistive load.

Section 1

6.1 Introduction

In this unit you will find that the emphasis of the course has changed. The first five units were primarily concerned with establishing the behaviour of electronic components by examining their *internal* structure. They discussed basic physical principles and used them to explain the workings of resistors, capacitors, inductors, transformers, diodes and transistors. From this unit on we will be concerned primarily with electronic components connected together to form what is called a *circuit* or *network*. We will concentrate on the voltages and currents associated with circuit elements (rather than internal fields) and will see how they are related to one another due to the connections between elements. For these purposes, the primary results from units 1 to 5 are the relationships between the current and voltage of each element.

circuit or network

Try the following revision SAQs.

SAQ 1

Write down the expressions for the voltage and current in (a) a resistor, (b) an inductor and (c) a capacitor.

SAQ 1

SAQ 2

What is the total resistance of two resistors R_1 and R_2 if they are connected (a) in series and (b) in parallel?

SAQ 2

We will develop a number of useful techniques which will enable you to determine any of the currents or voltages in a complex network (a subject called *network* or *circuit analysis*) and also enable you to determine the desired component values to give the required network behaviour.

All electronic networks can be thought of as having a common function: the processing or generation of changing voltages and currents. Whether the voltage or current in question is a train of logical pulses (as from a computer circuit), a VHF television signal, or a many kilowatt current through a power transmission line, there are a number of common techniques which can be used to determine the response of the network to it. When a varying voltage or current is used to convey information, as in communication or computation equipment, it is called a *signal* and the effect of an electric circuit upon it is called *signal processing*.

signal processing

With a knowledge of the internal structure of electronic components from the first five units, and a knowledge of the techniques of network analysis and signal processing given in the next four units, you will be ready for the design studies of the later units which discuss the application of electronics to build equipment to do particular jobs – amplifiers of different types, a dc power supply, a time base generator, a servomechanism and a curve tracer.

In this unit we will concentrate upon circuits consisting of a resistor plus (a) a capacitor, (b) an inductor and (c) both an inductor and a capacitor, and also the way they process such basic signals as a 'step' in voltage or current. These circuits are important because they are the basic building blocks out of which more complex circuits are made. Moreover, the form of the response and the methods of finding it are very similar to those of more elaborate circuits.

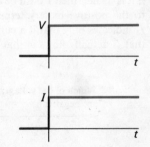

Figure 1 Voltage and current steps.

The home experiments for this unit are meant to illustrate the step response of three circuits described in the text and to give you some intuitive 'feel' for the signals on which to hang the mathematical descriptions in the text. You should not spend more than one hour on them.

If you are behind or have trouble with this unit, concentrate your efforts on section 6.2 covering Kirchhoff's laws (which are fundamental to all circuit analysis), and sections 6.4 and 6.5 covering the step response of *RC* and *RL* circuits.

Sections 6.6 and 6.7 are elaborations of the same step response ideas. Section 6.8 covers the more difficult step response of an *RLC* circuit. Place your emphasis on understanding the different types of step response that occur for different relative values of *R*. The mathematical details are there primarily for future reference.

6.1.1 Some network conventions and notation

Before we begin looking at the ways in which networks process signals we need to establish some notation and other conventions. You have already met the terms electric potential and emf, both of which are measured in volts. In discussing electronics from the point of view of circuits, any potential difference or emf is usually referred to as a *voltage*, without regard to how it was generated. On network diagrams it is usually denoted by the letter V, often with some subscript as in figure 2.

The arrow indicates the reference direction in which the voltage is measured or calculated. Thus V_3 is the potential at point E, (where the arrow points) measured with respect to point C. If the potentials at C and E with respect to earth are V_C and V_E respectively then

$$V_3 = V_E - V_C$$

V_3 can also be described as the *voltage drop* from E to C. Notice that a voltage *drop* is in the opposite direction to the reference arrow (which points from C to E for V_3).

Voltages are sometimes described using a double subscript. For example $V_3 = V_{EC}$, $V_2 = V_{CD}$. The first subscript indicates the point at which the voltage is measured, the second indicates the reference point.

voltage

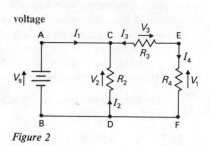

Figure 2

voltage drop

> Express the following voltages in figure 2 in terms of V_s, V_1, V_2 and V_3.
> (a) V_{DC}, (b) V_{CE}, (c) V_{AB}, (d) V_{EF}

(a) $V_{DC} = -V_2$, (b) $V_{CE} = -V_3$
(c) $V_{AB} = V_s$, (d) $V_{EF} = V_1$

The currents in a circuit are denoted by the letter *I*. The arrow indicates the reference direction in which currents are measured or calculated. The current at any point in a circuit is the rate at which positive charge passes that point in the reference direction, or at which negative charge passes in the opposite direction. Since an electron has a negative charge, the physical direction of flow of an electron current will be opposite to the direction in which a meter will indicate. However, the physical direction of flow is rarely of importance in circuit analysis.

The *reference* direction of a current or voltage should not be confused with the *sign* of that current or voltage. The reference arrow for V_3 does not mean that the potential of point E is actually higher than that of point C. It simply means that if it is higher, then V_3 will be a positive voltage. In fact, since point C is connected to the positive battery terminal, V_3 is negative. Similarly, the arrowhead for I_3 indicates only that if a current flows from E to C, I_3 will be positive, and not that I_3 actually flows in that direction. In fact, I_3 is negative in figure 2.

> Which of the voltages and currents indicated in figure 2 are positive and which are negative?

I_1, I_4, V_1 and V_2 are positive.
I_2, I_3 and V_3 are negative.

When beginning to analyse a circuit we usually assign labels to all the voltages and currents we will need. Because of the conventions just described, it is not necessary to determine the actual directions of current flow or voltage drop in advance. Any currents or voltages whose directions are opposite to those we arbitrarily assigned will be found to be negative.

It is sometimes useful to indicate which voltages and currents are constant and which are varying in time. If so, it is conventional to use capital letters for constant quantities and lower case for time varying quantities.

letter convention
capital = constant
lower case = time varying

As explained in Unit 1, in diagrams like figure 2 it is assumed that the wires are ideal conductors and therefore have no voltage drop in them.

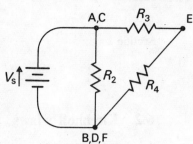

Figure 3 *A circuit which is electrically equivalent to the circuit in figure 2.*

> What does this mean about the length and shape of the wires as represented in the diagram?

Since the voltage drop is zero, the wires can be drawn to any length or shape which is convenient for the purposes of the diagram.

The point at which two or more wires are connected is called a *node*. The element plus wires joining any two nodes is called a *branch*. Although points B, D, and F in figure 2 are drawn separately, they actually represent a single node. This can be seen more clearly, if the circuit is redrawn as in figure 3. The two circuits are electrically identical.

node
branch

The source of voltage used in figures 2 and 3 is a battery. Actually, many other types of voltage source are used in electronics. (You will learn how to build one type, a dc power supply, in Unit 10.) The general symbol used to denote an *ideal voltage source* is shown in figure 4. The voltage of a source may either be constant (a dc source) as in figure 4(a), or may vary with time, as indicated by figures 4(b), (c) or (d). By an *ideal* voltage source, I mean that the voltage drop across it does not change no matter what current is drawn.

ideal voltage source

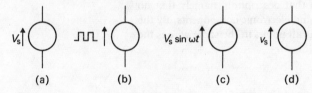

(a)   (b)   (c)   (d)

Figure 4 *Symbols for ideal voltage sources. (a) constant; (b) square wave; (c) sinusoidal; (d) general time varying.*

Is a battery an ideal voltage source?

No, as you will remember from Unit 2, if current is drawn from a battery, its terminal voltage decreases. The decrease gets larger, the more current is drawn. This is due to the internal resistance of the battery. No physical source of voltage is ideal and thus a representation of any real source must include a symbol for its internal resistance as well as the symbol for an ideal source when the effect of current drawn on its output voltage is being considered.

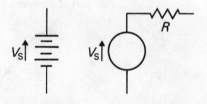

Figure 5 *A battery represented as an ideal voltage source in series with a resistor.*

In addition to sources of voltage, there are sources of current. An *ideal current source* provides a current which is the same no matter what voltage appears across its terminals. As with real voltage sources, any real current source has an internal resistance which must be included in any accurate representation of it. The symbol for a current source is shown in figure 6.

ideal current source

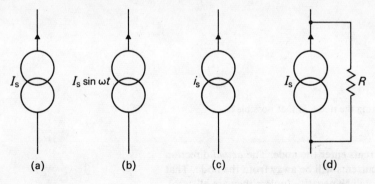

(a)   (b)   (c)   (d)

Figure 6 *Symbols for ideal current sources. (a) constant; (b) sinusoidal; (c) general time varying; (d) a representation of a real current source.*

6.2 Kirchhoff's laws

There are two different sets of relationships between currents and voltages which, jointly, determine their values in a circuit. There is the set of *internal* relations of the voltage and current at the terminals of each device, determined by the physics of the device. There is also a set of *external* relations between the voltages and currents of different devices, determined by the way in which the devices are interconnected. For a given circuit, these external relationships are found from two basic laws, Kirchhoff's current law and Kirchhoff's voltage law.

6.2.1 Kirchhoff's current law

At a node, a point at which several wires join, electrons enter along some of the wires and leave along others. The total number of electrons flowing towards the node will equal the number flowing away at any instant of time*, and so the sum of the currents entering the node must equal the sum of the currents leaving it.

There is an important assumption hidden in that description, namely that no current flows anywhere but in the conductors and electronic components. By the same principle, we conclude that the current at all points in the same wire is the same.

Apply the principle just described to the node in figure 7. (Assume that it is only part of a network.)

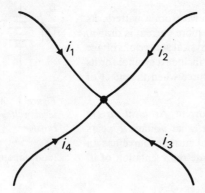

Figure 7

It gives $i_1 + i_2 + i_3 + i_4 = 0$.

All the currents are shown entering the node. Is that possible?

The *reference* directions of all four currents enter the node. The *actual* direction of at least one, and up to three of the currents will be away from the node. That is, some of the numbers i_1, i_2, i_3 and i_4 will be negative (unless they are all zero).

Kirchhoff's current law states that the sum of the currents entering any node is zero at all times.

Kirchhoff's current law

---

* *Strictly speaking, because of the thermal motion of electrons the numbers are equal only if averaged over a short time and not at 'any instant' of time. However, for most circuit analysis current is treated as though it were a smooth flow of charge. If it is necessary to include the effects of random thermal fluctuations, they are considered separately.*

A current i, which according to its reference direction leaves the node, is treated as a current of $-i$ entering the node.

> Notice that the statement of Kirchhoff's current law doesn't mention resistors, capacitors, etc., or alternating or direct currents. It simply tells how currents in connected wires are related.

Write down Kirchhoff's current law for nodes A, B and C in the circuit of figure 8. (I have used lower case letters to indicate currents because in circuits containing capacitors and inductors we are usually interested in varying voltages and currents.)

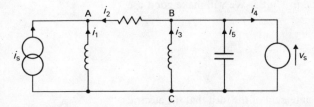

Figure 8

Node A

$$i_s + i_1 + i_2 = 0$$

Node B

$$-i_2 + i_3 + i_5 - i_4 = 0$$

Note the negative signs for i_2 and i_4. Note also that the unlabelled node to the right of B is really the same node as B, so far as any calculations are concerned. Similarly, the nodes adjacent to C are really part of C.

Node C

$$-i_s - i_1 - i_3 - i_5 + i_4 = 0$$

Figure 8 can be redrawn as in figure 9 with no change in its properties.

6.2.2 An extension of Kirchhoff's current law

Apply Kirchhoff's current law to all the nodes to the right of the dotted line in figure 10. You should have four equations. Add them. What can you conclude about the currents in elements crossed by the dotted line?

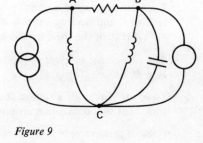

Figure 9

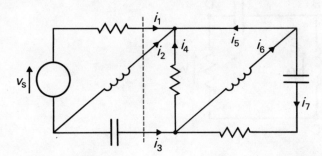

Figure 10

The four equations added together give

$$i_1 + i_2 + i_3 = 0$$

which is the sum of the currents entering the part of the circuit to the right of the dotted line.

Each of the remaining currents cancelled out because they entered one node and left another node to which Kirchhoff's current law was applied.

The dotted line divides the circuit into two parts, i.e. if all the branches which cross the dotted line are cut, the circuit will be separated into two disjoint parts. Thus Kirchhoff's current law applies not only to a node but to a part of a circuit.

The sum of all the currents entering any part of a network is zero at all times.

As when applying Kirchhoff's current law, care must be taken so that the proper sign is used for each current. It is useful to imagine the dotted line separating the part of the network from the rest. It is then clear which currents 'enter' and which 'leave'.

This method can be used to provide useful information even where a node cannot really be seen. In the transistor circuit of figure 11 we can apply Kirchhoff's current law to get $i_1 + i_2 + i_3 = 0$, i.e. the sum of the currents entering the transistor is zero.

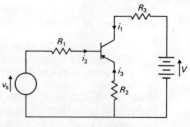

Figure 11

This important relationship already appeared in Unit 5. We will make good use of it later in the course.

6.2.3 Kirchhoff's voltage law

Kirchhoff's voltage law (KVL) is really a restatement of the idea that the electric potential between two points is the same along any two paths between them. (If you do not remember why the 'path' is involved reread the definition of potential in Unit 1.) In figure 12 the voltage drop from point A to point C is either V_s or, via B, $V_1 + V_2$.

Thus

$$V_s = V_1 + V_2$$

or

$$V_1 + V_2 + (-V_s) = 0$$

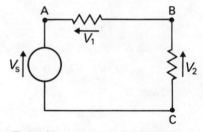

Figure 12

The latter expression is the sum of the voltage drops around the 'loop' from A to B to C and back again to A. (The voltage drop of the source from C to A is $-V_s$.) It is in the form in which Kirchhoff's voltage law is usually given.

The sum of the voltages (or voltage drops) around any closed path (or loop) is zero at all times.

Kirchhoff's voltage law

To apply Kirchhoff's voltage law correctly, the voltages must be added with a consistent sign convention around the loop.

Look at this example to see how it is done.

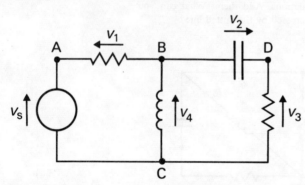

Figure 13

Apply Kirchhoff's voltage law along paths (a) ABCA, (b) ABDCA, (c) BCDB, and (d) DCBD.

(a) $v_1 + v_4 - v_s = 0$
(b) $v_1 - v_2 + v_3 - v_s = 0$
(c) $v_4 - v_3 + v_2 = 0$
(d) $v_3 - v_4 - v_2 = 0$

When applying Kirchhoff's voltage law which direction must we follow around the loop, and where must we start?

It doesn't matter at all. The equivalence of expressions (c) and (d) shows why.

How do different types of elements (resistors, capacitors, etc.) affect the form of the equations obtained by Kirchhoff's voltage law?

Not at all. Kirchhoff's voltage law is concerned with the way network elements are connected to each other, not with what these elements are. Applying Kirchhoff's voltage law to the circuit of figure 14 (where the nature of the element is left unspecified) gives exactly the same equations as found from figure 13.

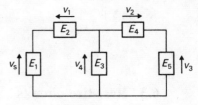

Figure 14

Exercise:

Figure 15 is a schematic representation of a circuit.

1. Indicate which of the following voltages *cannot* be determined by Kirchhoff's voltage law, from the voltages given. Determine the values of the others.

   V_{AH}, V_{BH}, V_{GH}, V_{FH}, V_{AD}, V_{CE}, V_{HD}

2. Find each of the following currents using Kirchhoff's current law.

   I_1, I_2, I_3, $I_4 + I_5$, $I_6 + I_7$

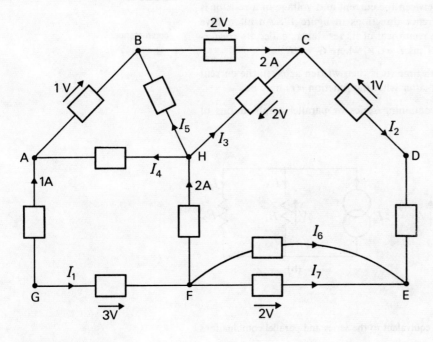

Figure 15

1. To determine each voltage we must find a path which makes a loop with the desired voltage and which contains only branches where voltages are given. Thus V_{GH}, V_{FH} and V_{CE} cannot be determined because there is no such path. The paths, and results for the other voltages are:

Path ABCH, $V_{AH} = -5\ V$;
Path BCH, $V_{BH} = -4\ V$;
Path ABCD, $V_{AD} = -2\ V$;
Path HCD, $V_{HD} = 3\ V$.

2. Applying Kirchhoff's current law to the nodes of the circuit in the following order gives:

node G, $I_1 = -1\ A$;
node F, $I_6 + I_7 = -3\ A$;
nodes D and E, $I_2 = 3\ A$;
node C, $I_3 = 1\ A$;
node H, $I_4 + I_5 = 1\ A$.

By applying Kirchhoff's current law to appropriately chosen parts of the network, rather than single nodes, any of these currents could be found in any order.

SAQ 3

In the circuit of figure 16, find V_1 and I_1.

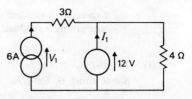

Figure 16

Section 3

6.3 Resistive circuits

Kirchhoff's laws are the basic relationships between voltages and currents in interconnected elements. We can now begin to apply them to different types of networks to see how these networks process electrical signals.

Let's start with the simplest combinations of components, voltage sources and resistors, and show that their behaviour with an arbitrary signal can be found from their response to a constant voltage. Then we will see that this convenient property also applies to a much larger group of circuits, those which don't contain capacitors or inductors.

You have already met simple networks consisting of sources of voltage and resistors in Unit 1. The simplest of all is that shown in figure 17.

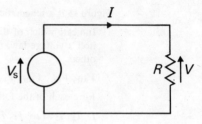

Figure 17

Remember that the relationship between the current and voltage in a resistor is $V = IR$ if V and I have the reference directions in figure 17. An alternative relationship for a resistor uses the reciprocal of its resistance, called its *conductance*. It is usually denoted by G. Thus $I = GV$, where $G = 1/R$.

conductance

Clearly, if the voltage changes with time (it is then written as $v(t)$), the current will be given by $i(t) = v(t)/R$, no matter what the function $v(t)$ may be.

Next in complexity are circuits containing series or parallel combinations of resistors.

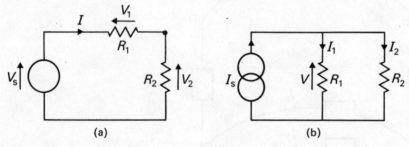

Figure 18

What resistance values are equivalent to the series and parallel combinations of resistors in figure 18?

(i) *Series*
$$R_s = R_1 + R_2$$

(ii) *Parallel*
If R_P represents the parallel combination of R_1 and R_2 then
$$\frac{1}{R_P} = \frac{1}{R_1} + \frac{1}{R_2}$$
so
$$R_P = \frac{1}{1/R_1 + 1/R_2}$$
$$= \frac{R_1 R_2}{R_1 + R_2}$$

Express the equivalent of a series and parallel combination of two resistors in terms of their conductances G_1 and G_2.

(i) *Parallel*
Substituting G_P for $1/R_P$, G_1 for $1/R_1$, and G_2 for $1/R_2$ gives
$$G_P = G_1 + G_2$$

(ii) *Series*

Substituting G_1, G_2 and $G_s = 1/R_s$ gives

$$\frac{1}{G_s} = \frac{1}{G_1} + \frac{1}{G_2}$$

or

$$G_s = \frac{G_1 G_2}{G_1 + G_2}$$

Find I, V_1 and V_2 in figure 18(a) and I_1, I_2 and V in figure 18(b). (Figure 18(a) is the same as the potentiometer circuit in Unit 1.)

(a) By Kirchhoff's current law applied to node A the current in both resistors is the same

$$I = \frac{V_1}{R_1}$$

$$= \frac{V_2}{R_2}$$

so

$$V_2 = \frac{R_2}{R_1} V_1$$

By Kirchhoff's voltage law

$$V_s = V_1 + V_2$$

$$= V_1 + \frac{R_2}{R_1} V_1$$

so

$$V_1 = \frac{R_1}{R_1 + R_2} V_s$$

similarly

$$V_2 = \frac{R_2}{R_1 + R_2} V_s$$

and then

$$I = \frac{V_s}{R_1 + R_2}$$

(b) By Kirchhoff's voltage law the voltage in the two resistors is the same

$$V = R_1 I_1$$

$$= R_2 I_2$$

so

$$I_2 = \frac{R_1}{R_2} I_1$$

By Kirchhoff's current law

$$I_s = I_1 + I_2$$

$$= I_1 + \frac{R_1}{R_2} I_1$$

so

$$I_1 = \frac{R_2}{R_1 + R_2} I_s$$

similarly

$$I_2 = \frac{R_1}{R_1 + R_2} I_s$$

and

$$V = \frac{R_1 R_2}{R_1 + R_2} I_s$$

Express these results in terms of the conductances of the resistors.

(a) $\quad V_1 = \dfrac{G_2}{G_1 + G_2} V_s \qquad V_2 = \dfrac{G_1}{G_1 + G_2} V_s \qquad I = \dfrac{G_1 G_2}{G_1 + G_2} V_s$

(b) $\quad I_1 = \dfrac{G_1}{G_1 + G_2} I_s \qquad I_2 = \dfrac{G_2}{G_1 + G_2} I_s \qquad V = \dfrac{I_s}{G_1 + G_2}$

The expressions for V_1 and V_2 illustrate the *voltage divider rule*:

voltage divider rule

A voltage across two series resistors splits so that the fraction of the total voltage across each resistor equals the ratio of that resistance to the sum of the two resistances.

The expressions for I_1 and I_2 illustrate the *current divider* rule:

current divider rule

A current through two parallel resistors splits so that the fraction of the total current through each resistor equals the ratio of the other resistance to the sum of the two resistances.

These two rules can be used together to simplify the analysis of networks, because they enable currents and voltages to be found without the explicit use of Kirchhoff's laws.

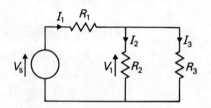

Figure 19

Let us use them to find the voltages and currents in the circuit in figure 19.

The first step is to combine R_2 and R_3 to get an equivalent resistance of $R_2 R_3/(R_2 + R_3)$. This is in series with R_1 so all three resistors are equivalent to one of magnitude $R_1 + R_2 R_3/(R_2 + R_3)$. Figure 20 shows the circuits at each step. Thus

$$I_1 = \frac{V_s}{R_1 + R_2 R_3/(R_2 + R_3)}$$

By the voltage divider rule

$$V_1 = \frac{R'}{R_1 + R'} V_s$$

$$V_1 = V_s \frac{R_2 R_3/(R_2 + R_3)}{R_1 + R_2 R_3/(R_2 + R_3)}$$

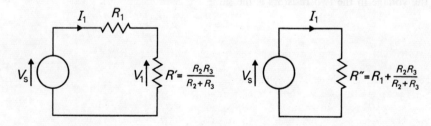

Figure 20   Circuits formed by combining the resistors in figure 19.

By the current divider rule

$$I_2 = \frac{R_3}{R_2 + R_3} I_1 = \frac{R_3}{R_2 + R_3} \times \left\{ \frac{V_s}{R_1 + R_2 R_3/(R_2 + R_3)} \right\}$$

In each of these examples all the voltages and currents we found were simply equal to the voltage or current of the source multiplied by some constant found from the resistance values. The sources used were constant sources.

Now suppose we replace the constant sources by time varying sources. All the steps used to find the expressions for the circuit's response to constant sources are still valid for varying sources. So to find the responses to a time varying voltage source v_s or current source i_s *no matter what signals v_s and i_s represent*, all that is required is the trivial step of replacing V_s by v_s or I_s by i_s in the expressions already found. This can be done because the current in a resistor *at any instant of time* equals the voltage *at that instant of time* multiplied by the resistance.

If you know the voltage across a capacitor at a given instant of time, can you find its current at that instant?

No. You need to know how that voltage is *changing*. Remember that the current in a capacitor is given by $i = C\,dv/dt$. Similarly, to determine the voltage across an inductor at some instant you need to know how the current through that inductor is changing, *not* its value at that instant. The voltage across an inductor is given by $v = L\,di/dt$.

Therefore, in a circuit containing capacitors and inductors you need to know something about the *form* of the source signals, the way in which they change in time, in order to find the other voltages and currents. For a network containing only resistors, as you saw above, we were able to find this information *without* knowing the form of the source signal. There are other electronic elements besides the resistors we have been considering which behave in this way. Any such component is called *resistive*.

A *resistive component* has the property that the currents and voltages at its terminals at any instant of time can be determined entirely from the applied voltages and currents at that instant.

resistive component

Capacitors, inductors and transformers do not have this property. They are called *reactive components*. Any circuit composed entirely of resistive components is a *resistive circuit*. As you might expect from the basic property of resistive circuits, the mathematical problem of finding all the voltages and currents in a resistive circuit is quite different from that in a circuit which also contains reactive components. Such circuits give rise to differential equations (see the mathematics booklet for a brief introduction to them) while resistive networks give rise to the more familiar algebraic equations. The solution of differential equations is inherently more difficult than the solution of algebraic equations. Thus techniques have been developed to enable reactive networks to be treated in a similar way to resistive networks. These techniques are some of the main topics of Unit 7. It is also why the study of resistive networks is so important.

reactive components
resistive circuit

Let us now look at some of the other electronic components which can be classified as resistive.

Strictly speaking, no device or circuit is truly resistive, just as no source of voltage is ideal. Any wire acts like a small inductor and any pair of wires forms a small capacitor. Similarly there is a small amount of inductance associated with all electronic components. These small reactances are called *parasitics* and have a noticeable effect only on currents and voltages which are changing very rapidly, i.e. high frequency signals. However, in the applications we will consider in this course, signals will generally be varying slowly enough so that the parasitic reactances can be ignored. Thus I will not include the rather cumbersome qualification 'except at high enough frequencies' at every mention of resistive devices.

In Unit 5 you met pn junction diodes and Zener diodes. Their symbols, together with a small sketch to remind you of the relation between their voltage and current are shown in figure 21. From these curves, you can determine the voltage at any instant given the current at that instant, or vice versa. Thus they are resistive devices.

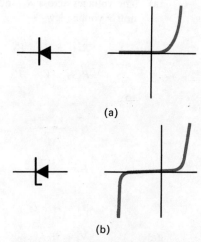

Figure 21 (a) *a* pn *junction diode;* (b) *a* Zener *diode and their v–i curves.*

Since a sketch of the V–I characteristic of a resistor would show just the straight line $V = IR$, a resistor is a *linear* resistive device. By contrast, the semiconductor devices are *nonlinear* resistive devices. In all cases there is a relationship between their current and voltage which depends only upon their instantaneous values. Diodes and resistors are two terminal devices. That is, each has two wire leads with which it is connected to other components. Thus there is only one voltage, and one current associated with them. Transistors are examples of devices which have three terminals. Thus they have three currents and voltages associated with them. These are shown in figure 22 with their usual reference directions. As with the two terminal devices, transistors are often described by their characteristic curves. However, these are usually *families* of curves as you saw in Unit 5.

linear and nonlinear resistive devices

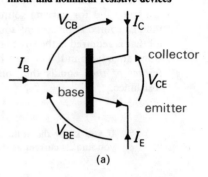

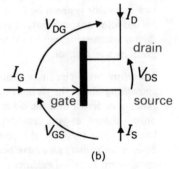

Figure 22 (a) Bipolar transistor; (b) field-effect transistor.

Can transistors be classified as resistive devices?

Yes, because the characteristic curves can be used to give some of the voltages and currents when others are given. It is not necessary to know more about any applied signal than its instantaneous value.

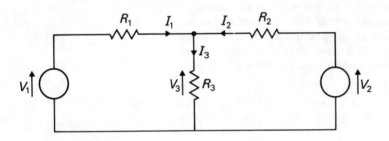

Figure 23

Exercise:

In the circuit in figure 23:

(a) Use Kirchhoff's voltage law to find I_1 and I_2 in terms of V_1, V_2 and V_3.
(b) Use Kirchhoff's current law to find I_3 in terms of I_1 and I_2.
(c) Use the results of (a) and (b) to find V_3 in terms of V_1 and V_2.
(d) If the constant voltage source V_1 is replaced by a sinusoidal voltage source $v_1 = V_1 \sin \omega t$ what voltage will appear across R_3?
(e) If the voltage source V_2 is replaced by a sinusoidal current source $i_2 = I_2 \sin \omega t$ (with the same sinusoidal source for v_1 as in (d)) what voltage will appear across R_3?

(a) The voltages across R_1 and R_2 are $I_1 R_1$, and $I_2 R_2$ respectively. By Kirchhoff's voltage law

$$R_1 I_1 = V_1 - V_3$$

$$R_2 I_2 = V_2 - V_3$$

so

$$I_1 = \frac{V_1 - V_3}{R_1}$$

$$I_2 = \frac{V_2 - V_3}{R_2}$$

(b) $\quad I_3 = I_1 + I_2$

(c) $\quad V_3 = R_3 I_3$
$\qquad = R_3(I_1 + I_2)$

Using the results of (a)

$$V_3 = \frac{V_1 R_2 R_3 + V_2 R_1 R_3}{R_1 R_2 + R_2 R_3 + R_1 R_3}$$

(d) Replacing V_1 by $v_1 = V_1 \sin \omega t$ in the expression found in (c) gives

$$v_3 = \frac{R_2 R_3 V_1 \sin \omega t + V_2 R_1 R_3}{R_1 R_2 + R_2 R_3 + R_1 R_3}$$

(e) From (a)

$$i_1 = \frac{V_1 \sin \omega t - v_3}{R_1}$$

but now $i_2 = I_2 \sin \omega t$. From (c)

$$v_3 = R_3(i_1 + i_2)$$
$$= R_3 \left(\frac{V_1 \sin \omega t - v_3}{R_1} + I_2 \sin \omega t \right)$$

Solving for v_3 gives

$$v_3 = \frac{(V_1 R_3 + I_2 R_3 R_1) \sin \omega t}{R_1 + R_3}$$

SAQ 4

The network shown in figure 24 is made up of resistive components. Figure 25 shows some hypothetical input and output waveforms.

1 Suppose $v_{out(1)}$ is the response of the network to $v_{in(1)}$. What is its response to $v_{in(2)}$?

2 Could $v_{out(2)}$ be the response to $v_{in(1)}$? If your answer is yes find the corresponding response to $v_{in(2)}$. If your answer is no, explain why not.

SAQ 4

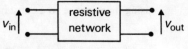

Figure 24

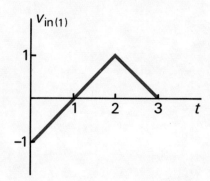
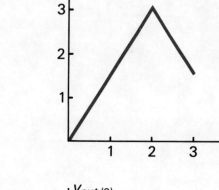
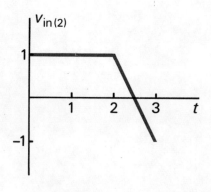
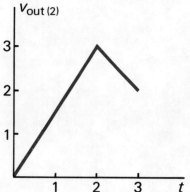

Figure 25 Waveforms for SAQ 4.

6.3.1 Summary of sections 6.2 and 6.3

The relationships between the currents and voltages of interconnected components are given by Kirchhoff's laws.

Kirchhoff's current law. The sum of the currents entering a node is zero at all times.

Kirchhoff's voltage law. The sum of the voltages (or voltage drops) around any loop is zero at all times.

Voltage divider rule

$$V_1 = \frac{R_1}{R_1 + R_2} V_s$$

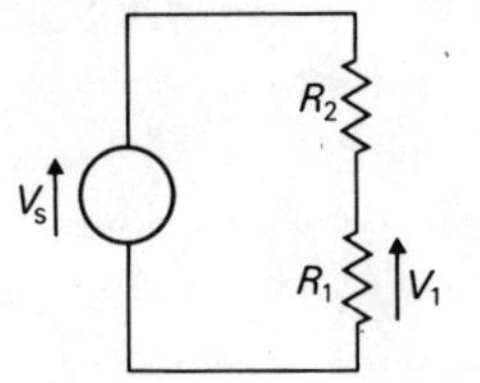

Current divider rule

$$I_1 = \frac{R_2}{R_1 + R_2} I_s$$

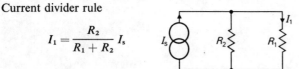

Figure 26

The voltages and currents at the terminals of a resistive component can be determined entirely from the applied voltages and currents at that instant. Therefore, the response of a resistive network to a signal which varies can be found using its response to a constant source.

Section 4

6.4 Resistance–capacitance circuits

The relative simplicity of the problem of signal processing in resistive networks is more easily appreciated by contrast with even very simple circuits containing capacitors or inductors. Let us start by considering the circuit in figure 27. The television programme for this week looks at this circuit at length and should give you some understanding of it.

As with our discussion of resistive circuits, we start by considering what happens if the voltage source is a battery or some other constant source, and by assuming that all other currents and voltages in the circuit are constant as well.

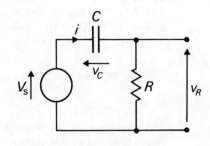

Figure 27 An RC circuit.

Remember that a capacitor consists of two conducting plates separated by an insulator. The current 'through' a capacitor is due to a change in the charge accumulated on its plates, rather than an actual flow of electrons from one plate to the other. With a constant voltage across it, the capacitor charge is also constant (because $Q = CV$) and so the current is zero. The same conclusion follows directly from the capacitor equation

$$i = C \frac{dv_C}{dt}$$

since the rate of change of the capacitor voltage dv_C/dt is zero. With no current flowing the resistor voltage v_R is zero and so the capacitor voltage v_C must equal the source voltage V_s no matter what that voltage is.

Next let us consider what happens when the voltage applied to the RC circuit is a voltage step rather than a constant. It is useful to think of a voltage step as generated either by a battery or a constant source with a switch, or by a hypothetical 'step generator' which produces v_s as shown in figure 28.

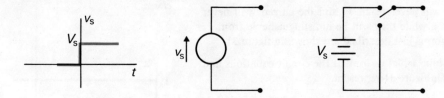

Figure 28 A voltage step can be generated by a step generator or by a battery and a switch.

If we tried to predict the response of the circuit to a voltage step as if it were a resistive circuit we would predict that the capacitor voltage would be zero before the switch closed and would jump instantly to V_s afterwards. The resistor voltage would remain zero throughout, as it was for all constant source voltages.

In fact the capacitor voltage does change from zero to V_s, but not instantly. Moreover, during the transition time the resistor voltage is not zero because there is a flow of current. Since current flows only during this transition period, the waveforms which occur during it are called *transients* and form the *transient response* of the circuit. This transient nature of the response of many circuits to a step of voltage or current supplied the name of this course unit.

transient response

In the first home experiment associated with this unit you will display and measure the transient response of an RC circuit on your home oscilloscope.

One simple way to see such fleeting behavior on an oscilloscope is to repeat it periodically and superimpose the responses which result*. For this reason you will use a square wave, which is simply a succession of positive and negative steps, to display the step response of the circuit.

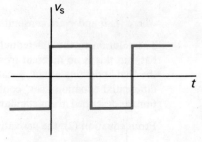

Figure 29 A square wave.

* A second way, used on this unit's TV programme, is to use a storage oscilloscope, whose trace does not fade after it is completed.

Let us now examine the voltages and currents which occur during the transition period more closely, first in a fairly intuitive manner and then by solving the equations of the circuit.

The key to understanding this circuit is an understanding of why the capacitor voltage doesn't change instantly to V_s when the switch is closed. Let's suppose it did change instantly. If we assume that the capacitor voltage was zero before the switch closed, then the accumulated charge will also change instantly from 0 to CV_s. But this change in accumulated charge in zero time constitutes an infinite current (for that instant only).

Why can't an infinite current flow?

The maximum current is limited by the resistor. The resistor voltage can't be larger than the source voltage V_s and so the largest current possible is V_s/R, which is certainly not infinite! In fact this same argument applies not only to a change in the capacitor voltage from 0 to V_s, but for *any* instantaneous change in its voltage.

The voltage on a capacitor cannot change instantaneously so long as there is resistance in the circuit to limit the current to a finite value.

The situation just after the switch is closed is shown in figure 30. The capacitor voltage is zero and so to satisfy Kirchhoff's voltage law, the resistor voltage is V_s. The current is V_s/R which begins to charge the capacitor.

Once the current begins to flow the capacitor voltage increases smoothly from zero. But then the resistor voltage must decrease, because it is the difference between the source and capacitor voltages. Thus the charging current $i\,(=v_R/R)$ also decreases. The capacitor voltage still increases, but more slowly. This process will continue with the capacitor voltage rising towards V_s at an ever decreasing rate because as it gets closer the resistor voltage, and therefore the charging current, gets smaller.

In theory the capacitor voltage will never reach V_s and the current will never reach zero but, in practice, after a while they will be indistinguishable from V_s and zero respectively. The waveforms just described are shown in figure 31.

To find out how long 'after a while' is, let us look at the circuit equations and find out exactly what the curves in figure 31 represent.

We apply Kirchhoff's voltage law to the circuit immediately after the switch is closed.

$$V_s = v_C + v_R \quad\text{or}\quad V_s = v_C + RC\frac{dv_C}{dt}$$
$$= v_C + Ri$$

or

$$RC\frac{dv_C}{dt} = -v_C + V_s \qquad (1)$$

This is a first order, linear differential equation. Its solution is a function of the form:

$$v_C = A\exp\frac{-t}{\tau} + B \qquad (2)$$

where A, B and τ are constants we must now evaluate.

The solution to the differential equation appears to have been pulled out of a hat, but this is no different from what you do when you multiply 7 by 6. There the 'hat' contains a multiplication table which you memorized as a child. The differential equation 'hat' contains the solutions to common differential equations as described in any standard introductory textbook on differential equations.

From equation (2), the derivative of v_C is

$$\frac{dv_C}{dt} = \frac{-A}{\tau}\exp\frac{-t}{\tau} \qquad (3)$$

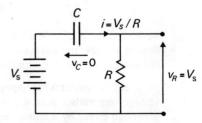

Figure 30 *The RC circuit with voltages indicated as when the switch has just closed.*

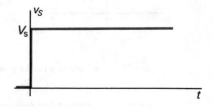

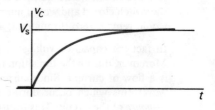

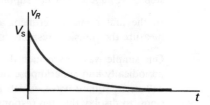

Figure 31 *The transient waveforms in an RC circuit.*

If we substitute these expressions for v_C and dv_C/dt back into our original equation (1), we get

$$-\frac{RC}{\tau} A \exp \frac{-t}{\tau} = -A \exp \frac{-t}{\tau} + B - V_s \tag{4}$$

Equation (4) should hold for all values of t. To do this the right and left-hand sides must be the same. This means that:

1 The coefficients of the exponential terms on both sides must be the same. Thus

$$\frac{RC}{\tau} = 1 \quad \text{or} \quad \tau = RC$$

2 There is a constant term on the right but not on the left. Thus

$$0 = B - V_s \quad \text{or} \quad B = V_s$$

To determine A we use the fact that the capacitor voltage is zero immediately after the switch is closed. As explained in the mathematics booklet, to solve a differential equation we always need to know some information in addition to the equation itself, usually a particular value of the solution or its derivative at one point in time.

With $t = 0$, equation (2) becomes

$$v_C(0) = 0$$
$$= A \exp 0 + V_s$$

so

$$A = -V_s$$

Now that we have found the values of the three constants A, B and τ the complete solution becomes

$$v_C = -V_s \exp \frac{-t}{RC} + V_s \tag{5}$$

or

$$v_C = V_s \left(1 - \exp \frac{-t}{RC}\right)$$

expression for capacitor voltage

and, using Kirchhoff's voltage law

$$v_R = V_s - v_C = V_s \exp \frac{-t}{RC} \tag{6}$$

Equations (5) and (6) are the mathematical expressions for the curves shown in figure 31. Both are exponential functions which start at a given value at $t = 0$ and move exponentially towards their final values.

The constant $\tau \, (= RC)$ is called the *time constant*, and determines how quickly v_R and v_C approach their final values. To see this we can look at v_C for several values of t.

After a time equal in duration to one time constant, v_C has increased to 63% of its final value. By three time constants it is up to 95%, and to 99.3% by five time constants.

At the end of *any* interval of duration τ, v_C will have increased by 63% of the *difference* between its value at the beginning of that interval and its final value. Although it will never reach its final value, for many practical purposes v_C can be considered to be equal to its final value after a small number (say three to five) of time constants.

Similarly v_R covers the same percentage of the change from its initial to its final value, zero, after corresponding multiples of the time constant.

Study comment

> This is an appropriate time to try Home Experiment 1, which examines the step response of the *RC* circuit.

Table 1

t	v_C
τ	$0.633 V_s$
2τ	$0.865 V_s$
3τ	$0.950 V_s$
4τ	$0.982 V_s$
5τ	$0.993 V_s$

6.5 Resistance–inductance circuits

Now let us look at the step response of a different circuit, shown in figure 32. Here the switch *opens* at $t=0$ to give a current step.* We apply Kirchhoff's current law to node A.

$$I_s = i_R + i_L$$

$$= \frac{v_R}{R} + i_L$$

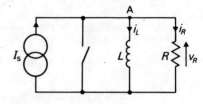

Figure 32 *An RL circuit.*

But since

$$v_R = L \frac{di_L}{dt}$$

$$I_s = \frac{L}{R} \frac{di_L}{dt} + i_L$$

or

$$\frac{L}{R} \frac{di_L}{dt} = -i_L + I_s \qquad (7)$$

Equation (7) has exactly the same form as equation (1). Only the variable names are different. Thus we needn't go through the entire solution.

Try writing down the solution to equation (7) by analogy with equations (1) and (5).

If you replace v_C with i_L, V_s with I_s and RC with L/R in equation (5) you get the solution to equation (7).

$$i_L = -I_s \exp \frac{-tR}{L} + I_s$$

$$i_L = I_s \left(1 - \exp \frac{-tR}{L}\right)$$

expression for inductor current

We can now find v_R from i_L by using Kirchhoff's current law to find the resistor current:

$$i_R = I_s - i_L$$

$$= I_s \exp \frac{-tR}{L}$$

$$v_R = R i_R$$

$$= R I_s \exp \frac{-tR}{L}$$

---

* *Notice that before the switch opens the current generator is short circuited. The full current I then flows through the short. Note that we cannot use a switch in series with the current source (as with the voltage source), since the current generator is a device which is designed to maintain a constant current I, but an open switch in series would force the current to be zero. In theory, this is a contradictory model. In practice, it might mean a burnt out current generator. In the same way a voltage source cannot have a switch across it.*

The waveforms found appear as in figure 33. They are exponential functions again. The time constant here is $\tau = L/R$.

Now let us look at the step response of the RL circuit from a physical point of view. You should notice the similarities between the following discussion and that describing the step response of the RC circuit.

Following an instantaneous change in the source current from zero to I_s, the inductor current also changed from zero to I_s, but exponentially rather than instantly. The key to the RL circuit is an understanding of why the inductor current doesn't change instantly.

Suppose the current did instantly jump from zero to I_s. Since the current is proportional to the flux, the flux will also jump instantly. According to Faraday's law, a change in flux in zero time will generate an infinite voltage across the inductor. The resistor is directly across the inductor so its voltage will be infinite too. But an infinite voltage across a resistor demands an infinite current through it. The maximum current which can flow through the resistor is the total source current I_s, which is not infinite. So, due to the resistor, the inductor current cannot change instantly.

The current in an inductor cannot change instantaneously so long as there is resistance in the circuit to limit its voltage to a finite value.

The situation immediately after the switch closes is shown in figure 34. The inductor current is still zero so the entire source current I_s flows through the resistor. Thus the resistor voltage (and the inductor voltage) is $I_s R$. From the inductor equation $v_R = L\, di_L/dt$, if the inductor voltage is positive the inductor current will increase from zero. Any current in the inductor must be subtracted from the current in the resistor since the total current from the source remains a constant I_s. But if the resistor current decreases, so must the resistor voltage. Thus the inductor current increases at a slower rate. If you look back at the description of the build up of the capacitor voltage in the RC circuit you will find that the situations are analogous. Hence we get the same exponential build up in both cases.

The similarities of the behaviour of the RC and RL circuits are worth pursuing a little further. As with the analogies drawn between electric and magnetic fields in your earlier course units, we can draw similar analogies between circuits which contain capacitors and inductors – electric and magnetic devices respectively. These analogies can be improved still further if in the RL circuit we use the alternate version of the resistor equation, $I = GV$.

The list of these analogies is given in Table 2.

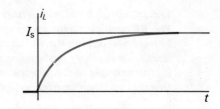

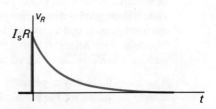

Figure 33 The transient waveforms in an RL circuit.

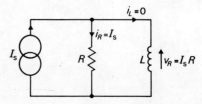

Figure 34 The RL circuit with voltages and currents indicated as when the switch has just opened.

Table 2

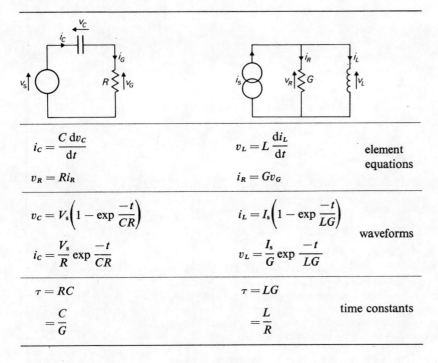

$i_c = \dfrac{C\, dv_c}{dt}$	$v_L = L\, \dfrac{di_L}{dt}$	element equations
$v_R = Ri_R$	$i_R = Gv_G$	
$v_c = V_s\left(1 - \exp\dfrac{-t}{CR}\right)$	$i_L = I_s\left(1 - \exp\dfrac{-t}{LG}\right)$	waveforms
$i_c = \dfrac{V_s}{R}\exp\dfrac{-t}{CR}$	$v_L = \dfrac{I_s}{G}\exp\dfrac{-t}{LG}$	
$\tau = RC$ $= \dfrac{C}{G}$	$\tau = LG$ $= \dfrac{L}{R}$	time constants

The *RC* circuit contains a *voltage* source in *series* with the other two elements. The *RL* circuit contains a *current* source in *parallel* with the other two elements. For every *voltage* in the *RC* circuit there is a corresponding *current* in the *RL* circuit. This analogy can be taken much further than these two particular circuits. It is an example of the principle of *duality* and comes from the basic laws of the elements and their interconnection.

duality

Given any circuit containing sources, resistors, capacitors and inductors, you can construct another circuit model with identical behaviour if you replace (a) parallel connections with series connections and vice versa, (b) voltage sources with currents sources and vice versa, (c) resistances with conductances and vice versa, (d) capacitors with inductors and vice versa, and (e) voltage waveforms with current waveforms and vice versa.

The importance of this principle is that the properties of a circuit can be found simply by analogy if the properties of its dual circuit are known.

Section 6

6.6 Other single time constant circuits

There are many other circuits with similar waveforms to those of the RC and RL circuits. Let us look at one with more components than the previous two circuits. From it we will try to extract some general rules to enable us to analyse any such circuit. Consider the circuit in figure 35. Assume that the switch closes at $t=0$ and that any transient waveforms from connecting the 2 V source have died down to a point where they can be neglected. Our interest is not only in the waveforms which will result from closing the switch, but in the methods used in setting up the equation. We start by applying Kirchhoff's current law to node A

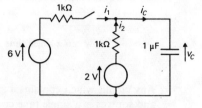

Figure 35

$$i_C = i_1 + i_2$$

Now using the capacitor equation

$$i_C = C \frac{dv_C}{dt} = 10^{-6} \times \frac{dv_C}{dt}$$

so

$$10^{-6} \times \frac{dv_C}{dt} = i_1 + i_2 \qquad (8)$$

We now want to express i_1 and i_2 in terms of v_C so that v_C is the only variable in the differential equation

$$1000 \times i_1 = 6 - v_C$$

$$1000 \times i_2 = 2 - v_C$$

Can you see the loops to which we applied Kirchhoff's voltage law to get these two expressions?

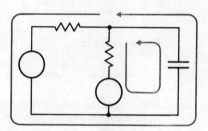

If we solve these two expressions for i_1 and i_2 and substitute them into equation (8) we get

$$10^{-6} \times \frac{dv_C}{dt} = \frac{6 - v_C}{1000} + \frac{2 - v_C}{1000}$$

or

Figure 36 The loops used to find the equations for i_1 and i_2.

$$5 \times 10^{-4} \frac{dv_C}{dt} = -v_C + 4 \qquad (9)$$

Again we find a differential equation of the same form as before.

To find the complete solution of the RC circuit we needed the value of the capacitor voltage at $t=0$. What is it for this circuit?

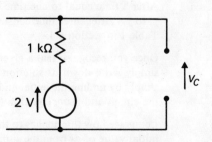

Since there is no path which will permit an infinite current to flow, the capacitor voltage cannot change instantly and so will be the same just before and just after the switch closes. We can find it most easily before the switch closes.

Since we assumed that all transient effects of connecting to 2 V source have died out, the capacitor appears like an open circuit before the switch is closed. With this fact and the open switch, all that remains of the circuit is what is shown in figure 37.

Figure 37 The circuit used to find the initial value of v_C.

No currents flow, so the voltage across the resistor is zero. Thus $v_C = 2$ V at the instant the switch is closed.

Notice that even though there are several resistors and two sources, the *form* of the differential equation is the same as in the *RC* and *RL* circuits. Thus its solution will have the same form as their solutions. Any circuit which has an equation, and hence solutions, of this same form is called a *single time constant circuit* because the time varying part of the step response is a single exponential.

single time constant circuit

How can a single time constant circuit be recognised?

Any circuit which contains a single capacitor or a single inductor plus resistors and sources is a single time constant circuit.

As a general rule, networks which contain more than one inductor or capacitor or both inductors and capacitors will not be single time constant circuits. However, there are some special cases in which a circuit might contain several capacitors (or several inductors) which are connected in such a way that they can be combined to form one equivalent capacitor (or inductor). In this case the network will still be a single time constant network with a time constant equal to the equivalent capacitance times the equivalent resistance (or equivalent inductance divided by the equivalent resistance).

Before going on to find the solutions of equation (9), let us look at the general equation and see if we can't write its solution in a more convenient form.

$$\tau \frac{dy}{dt} = -y + K \tag{10}$$

This is our differential equation. Notice that the constant multiplying the derivative is τ, which as we saw in the *RC* and *RL* circuit is the time constant of the solution. The constant K came from the sources in the two circuits. To solve the differential equation we also need an initial value of y which we denote by y_i. We have already seen that the solution is of the form

$$y = A \exp \frac{-t}{\tau} + B \tag{11}$$

Its derivative is

$$\frac{dy}{dt} = -\frac{A}{\tau} \exp \frac{-t}{\tau}$$

Notice that the derivative approaches zero as time goes to infinity so that, from the differential equation (10), the *final value* of y is the constant K. We denote this by y_f. But from the solution (equation (11)) the final value is B. Thus $y_f = K = B$.

The value of A is found from the initial value of y. From equation (11), at $t = 0$, $y = A + B$ or $A = y_i - y_f$.

Thus the complete solution is

$$y = (y_i - y_f) \exp \frac{-t}{\tau} + y_f \tag{12}$$

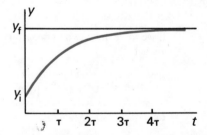

Figure 38 *The solution to the differential equation of a single time constant circuit.*

It is an exponential function, which starts at y_i at $t = 0$, and decays with a time constant of τ to y_f at $t = \infty$, as in figure 38.

general form of solution for single time constant circuit

After a time equal to one time constant it has covered 63% of the change from y_i to y_f. After three time constants it has covered 95%, etc., just as described in Table 1 in section 6.4.

Once you recognise that a given circuit is a single time constant circuit you can simply write down its solution *without even setting up the differential equation*, simply by finding the initial and final values and the time constant directly from the circuit and using equation (12).

Let us see how this applies to the circuit of figure 35. We have already found the initial value of v_C from the simplified circuit of figure 37.

What is the final value of v_C?

As t tends to ∞ the transient response decays to zero and only direct currents flow, so the capacitor is effectively an open circuit. We can now find the final value of v_C from another simplified circuit, figure 39(a).

By the voltage divider rule

$$V_1 = \frac{1000}{1000 + 1000}\{6 + (-2)\}$$
$$= 2\text{ V}$$

The final value of v_C is

$$v_C = V_1 + 2$$
$$= 4\text{ V}$$

All we need to find now is the time constant τ, which equals $C \times R$, where R is the total resistance which keeps the capacitor voltage from changing instantaneously. We need an equivalent resistance equal to the total resistance connected across the capacitor terminals. To find this we use a third simplified circuit, figure 39(b). Since we are only interested in a resistance, and since our ideal sources have no resistance,* we can set them to zero (for the purposes of this calculation only).

The two resistors in parallel are equivalent to a single resistance $R = 500\ \Omega$. Thus $\tau = CR = 0.5 \times 10^{-4}$.

The two constants we have just calculated directly from the circuit, the final value of the capacitor and the time constant, are the same constants that appear in the differential equation (9).

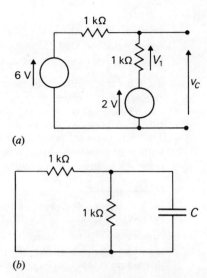

Figure 39(a) *The circuit used to find the final value of v_C; (b) the circuit used to find the equivalent resistance.*

Use these three constants and equation (12) to write down the expression for v_C.

$$v_C = 2\exp\frac{-t}{0.5 \times 10^{-4}} + 4$$

The initial and final values of the capacitor voltage and the time constant were found from three simplified circuits which were *purely resistive*. The reasoning we used to find these simplified circuits was general so we can now summarize it for use in any single time constant circuit.

The step applied to a single time constant circuit may be caused by one or more constant voltage or current sources and one or more switches, so long as they close or open simultaneously. It may also be caused by sources which themselves generate steps, such as a square wave or pulse generator. In any case there is one set of source values and/or switch settings before the step is applied and a second set after it is applied.

The final value of the capacitor voltage or inductor current is an equilibrium value. When it is reached no voltages or currents are changing. Thus a capacitor current is zero and an inductor voltage is zero.

To find the final value of a capacitor voltage, replace the capacitor by an open circuit and find the voltage at its terminals from the resulting circuit, with the sources or switch(es) set so that the step is applied.

rule for finding final values

To find the final value of an inductor current, replace the inductor by a short circuit and find the current *through this short circuit* from the resulting circuit with the sources or switch(es) set so that the step is applied.

The initial value of a capacitor voltage or inductor current may or may not be an equilibrium value. If it is we can find its value from a simplified circuit, relying on the fact that a capacitor voltage or inductor current will be the same the instant before and the instant after the step is applied.

* Here you can clearly see the advantage of separating a real source into two parts: its internal resistance plus an ideal source.

To find the initial value of the capacitor voltage, replace the capacitor by an open circuit and find the voltage at its terminals from the resulting circuit, with the sources or switch(es) set so that the step is not applied.

rule for finding initial values

To find the initial value of the inductor current, replace the inductor by a short circuit and find the current *through this short circuit* from the resulting circuit with the sources or switch(es) set so that the step is not applied.

The time constant is determined by the resistance which keeps the capacitor voltage or inductor current from changing instantaneously. It does not depend upon the sources in the circuit.

To find the time constant, find the equivalent resistance across the capacitor or inductor terminals from the circuit which results from setting all ideal sources to zero (that is, voltage sources become short circuits, current sources become open circuits) with the switch(es) set so that the step is applied. The time constant is then the capacitance multiplied by this equivalent resistance or the inductance divided by this equivalent resistance.

rule for finding time constants

Exercise:

In each of the following circuits find the initial and final values of the capacitor voltage or inductor current and the time constant. Calculate and sketch roughly the corresponding waveforms. Assume that at $t=0$ the switches *close* in (a) and (d) and *open* in (b) and (c).

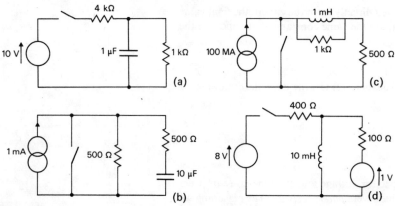

Figure 40

(a) With the switch open there is no source of voltage across the capacitor so the initial value is $v_C = 0$ V. With the switch closed and the capacitor replaced by an open circuit the final value is $v_C = 2$ V, by the voltage divider rule. With the switch closed and the source set to zero the two resistors are in parallel. They are equivalent to 800 Ω. Thus $\tau = 800\ \Omega \times 1\ \mu F = 0.8$ ms.

$$v_C = 2 - 2\exp\frac{-t}{0.8 \times 10^{-3}}.$$

(b) With the switch closed the initial value is $i_L = 0$ A. With the switch open and the inductor replaced by a short circuit the current through that short is the total source current whose final value is $i_L = 100$ mA. With the switch open and the current source replaced by an open circuit only the 1 kΩ resistor appears across the inductor. Thus

$$\tau = 1\ \text{mH}/1\ \text{k}\Omega = 1\ \mu s$$

$$i_L = 0.1 - 0.1\exp\frac{-t}{10^{-6}}$$

(c) Initially $v_C = 0$ V. With the switch open and the capacitor replaced by an open circuit no voltage appears across the resistor in series with the capacitor and thus the final value is $v_C = 500 \times 10^{-3} = 0.5$ V. The equivalent resistance is 1 kΩ so $\tau = 0.01$ s.

$$v_C = 0.5 - 0.5\exp\frac{-t}{0.01}$$

(d) With the switch open and the inductor replaced by a short circuit, $i_L = 1/100 = 10$ mA. With the switch closed and the inductor replaced by a short circuit, the current from the 8 V source also flows through the short. That current is 8/400 = 20 mA. The final value of the total inductor current is $i_L = 10 + 20 = 30$ mA. The two resistors in parallel form the equivalent resistance of 80 Ω so $\tau = 0.8$ s.

$$i_L = 0.03 - 0.02 \exp \frac{-t}{0.8}$$

Study comment

The following SAQ contains some teaching points about Home Experiment 6.1.

SAQ 5

(a) The switch in the circuit in figure 41 changes from position 1 to position 2 at $t = 0$. Find an expression for the voltage waveform which appears across the capacitor and draw a rough sketch of it. (Assume the switch was in position 1 for at least 1 minute before changing.)

(b) Now suppose the switch changes back to position 1 after 5 s. Find an expression for the capacitor voltage which results and include this waveform in the sketch for part (a).

(c) What voltage waveform will appear across the capacitor if the switch continues to change from position 1 to position 2 and vice versa every 5 s?

SAQ 5

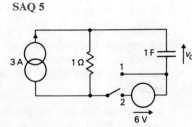

Figure 41

A correct answer to question (c) explains the waveform appearing in Home Experiment 6.1 in response to the 500 Hz square wave (with a different time scale, of course).

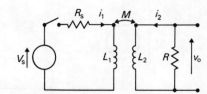

Figure 42 Transformer circuit with resistive load.

6.6.1 An unobvious example. A transformer with resistive load

As you are beginning to see, the step response we found for one particular and very simple RC circuit can be used as a basis for explaining the step response of a rather large variety of circuits. We will now look at one which contains a transformer rather than an inductor or a capacitor.

Let us find the response of the output voltage v_o in the network in figure 42 to a voltage step caused by closing the switch. We will make the common assumptions that the transformer is lossless and that the coils are perfectly coupled so that there is no leakage flux. The properties of a transformer with these assumptions were described in Unit 3, section 3.4.

To help us to analyse this circuit we introduce a simplified model of a transformer called an *ideal transformer*. Its symbol is shown in figure 43.

It is characterized by the very simple relationships

$$v_2 = N v_1$$
$$i_2 = -\frac{1}{N} i_1 \qquad (13)$$

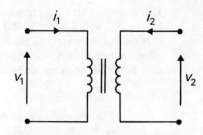

Figure 43 The symbol for an ideal transformer.

ideal transformer

This equation shows that the ideal transformer models the 'transforming' effect of a real transformer on its currents and voltages. The transformer in figure 42 can be represented by an inductor whose inductance represents the inductance of its primary winding in parallel with an ideal transformer as in figure 44. In this representation the *magnetizing current* flows through the shunt inductor and the remaining current flows through the ideal transformer. The primary, secondary and mutual inductances of an ideal transformer are infinite. Although this is not possible in practice, in many real transformers these inductances are large enough so that in a given application the magnetizing current is negligible compared to the total current. Then the ideal transformer alone is an adequate representation of the real transformer.

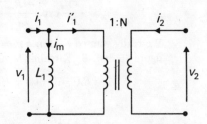

Figure 44 A transformer represented by an ideal transformer in parallel with an inductor.

Are you wondering how the circuit of figure 44 can represent the transformer without including its secondary or mutual inductances? If so try this exercise. Show that v_1 and v_2 in figure 44 can be expressed in terms of i_1 and i_2 by the equations below, which were shown in Unit 3 to be the equations of a transformer with perfect coupling. (Hint: start with $v_1 = L_1 \, di_m/dt$ and use $i_1 = i_m + i_1'$ and equation (13). Remember too that $N = (L_2/L_1)^{1/2}$ and $M = (L_1 L_2)^{1/2}$.)

$$v_1 = L_1 \frac{di_1}{dt} + M \frac{di_2}{dt}$$

$$v_2 = M \frac{di_1}{dt} + L_2 \frac{di_2}{dt}$$

Is an ideal transformer a reactive or a resistive device?

Resistive! Equation (13) shows that only instantaneous values are needed to find the currents and voltages in an ideal transformer. It is for this reason, and the simplicity of equation (13), that it is convenient to use an ideal transformer to model a real one at least to obtain a first order approximation to the overall behaviour.

Now let's get back to our problem. Using our new model, the circuit becomes as shown in figure 45.

Figure 45 The ideal transformer used in the circuit of figure 42.

Examine that part of the circuit to the right of the broken line. From equation (13), $i_1' = -Ni_2$ and $v_1 = v_0/N$. But $v_0 = -i_2 R$, so $i_1 = (N^2/R)v_1$. But this means that the entire network to the right of the broken line acts simply like a resistor of magnitude $R' = R/N^2$. Thus we can simplify the circuit, as in figure 46, which now looks like the kind we have learned to handle.

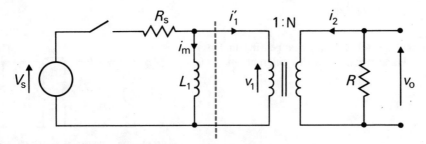

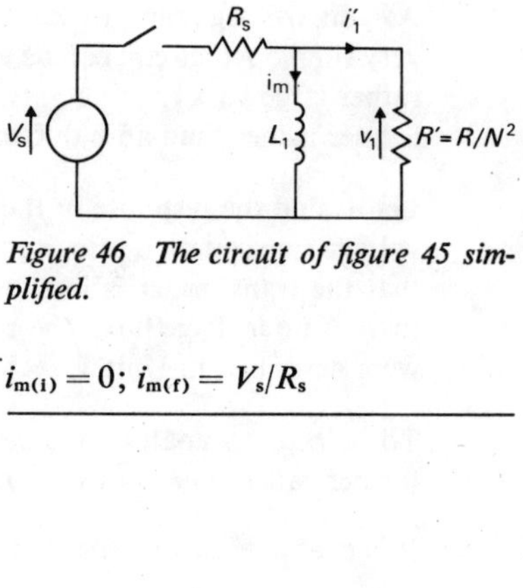

Figure 46 The circuit of figure 45 simplified.

Find the initial and final values of i_m and the equivalent resistance. Then find i_m using equation (12).

$i_{m(i)} = 0$; $i_{m(f)} = V_s/R_s$

The equivalent resistance is given by

$$R_{eq} = \frac{R_s R'}{R_s + R'}$$

$$= \frac{R_s R}{N^2 R_s + R}$$

$$i_m = \frac{-V_s}{R_s} \exp \frac{-t R_{eq}}{L_1} + \frac{V_s}{R_s}$$

We really wanted v_0. Since $v_0 = Nv_1$, we can find v_1 from the simpler circuit of figure 46 first. By Kirchhoff's voltage law

$$v_1 = R_s(-i_1' - i_m) + V_s$$

Notice that Kirchhoff's current law was quietly used to find the current in R_s. You should try to develop the ability to skip steps in this way by mentally applying Kirchhoff's laws when possible.

Now using $v_1 = R'i_1'$ we can eliminate i_1' and solve for v_1.

$$v_1 = \frac{R'}{R' + R_s}(V_s - R_s i_m)$$

Now putting all the pieces together and simplifying, we get

$$v_0 = \frac{R}{R + N^2 R_s} V_s \exp \frac{-t R_s R}{L_1(R + N^2 R_s)}$$

We have taken a circuit which, due to the transformer, was quite a bit more difficult than the simple RL circuit and reduced it to a series of several simpler problems.

6.6.2 Summary of sections 6.4, 6.5 and 6.6

These three sections described the step response of single time constant circuits. Any circuit which can be reduced to a single capacitor or a single inductor plus resistors and sources is a single time constant circuit.

The capacitor and resistor voltages for the RC circuit are

$$v_C = V_s\left(1 - \exp\frac{-t}{RC}\right)$$

$$v_R = V_s \exp\frac{-t}{RC}$$

The time constant of the RC circuit is $\tau = RC$.

The inductor current and resistor voltage for the RL circuit are

$$i_L = I_s\left(1 - \exp\frac{-tR}{L}\right)$$

$$v_R = RI_s \exp\frac{-tR}{L}$$

The time constant of the RL circuit is $\tau = L/R$.

The general expression for the capacitor voltage or inductor current in a single time constant circuit is

$$y = (y_i - y_f)\exp\frac{-t}{\tau} + y_f$$

where y_i is the initial value of y, y_f is the final value of y and τ is the time constant.

A signal which changes from an initial to a final value exponentially completes 63% of the transition after one time constant, 95% after three time constants and 99% after five time constants.

To find the expression for an exponential signal it is only necessary to find the initial and final values and the time constant, and then substitute into the general expression. The three constants can often be found from simplified models of the circuit.

The principle of duality describes an analogy between capacitors and inductors, currents and voltages, resistance and conductance, series and parallel connections. If one circuit is the dual of another then all properties of one circuit will be the same as the corresponding dual properties of the other.

Section 7

6.7 A nonlinear circuit. The dc restorer

In your home experiment, you applied a square wave to the simple *RC* circuit and should have found that the voltage across the resistor had an average value of zero. This average value remained zero even when a battery was added in series with the square wave generator.

In this section we will examine a similar circuit, but with a nonlinear resistive element, a diode, replacing the resistor in the earlier circuit. The circuit is shown in figure 47.

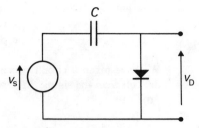

Figure 47 *A dc restorer circuit.*

dc restorer

The average value of v_D, the voltage across the diode, is not zero, as we shall see, and so the circuit is called a *dc restorer*. Because the diode is nonlinear, the differential equation of the dc restorer does not have the same form as those of the single time constant circuits we have discussed. Thus we cannot find the voltage waveform by directly applying our earlier results. However, by using a suitably simplified model of the diode, we can apply these results indirectly.

The simplifications in the diode model are great enough so that the theoretical waveforms resulting from it do not include several significant features of the waveforms in the real circuit, as you will see in your second home experiment. Nevertheless, the simplified model enables us to easily describe the main features of the circuit. It can then be refined to show some of the secondary features as indicated in SAQ 7 at the end of this section.

Remember that the basic property of a diode is that it presents much lower resistance values when it is forward biased than when it is reverse biased. So, as an approximation, let us assume that when it is forward biased it behaves as a linear resistance of magnitude R_F and when reverse biased, as a linear resistance of magnitude R_R. This gives it a *V–I* characteristic curve like that in figure 48.

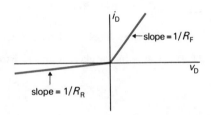

Figure 48 *An approximate characteristic for a diode.*

In this approximate model the transition from a high to a low resistance occurs sharply at zero volts. As you saw in Unit 5, in a real diode the *V–I* characteristic curve has no sharp corner so the transition from high to low resistance occurs over a range of voltages. Keep this in mind when you compare the results of your second home experiment with the theoretical results to be described now.

Suppose that the values of the forward and reverse resistances are 100 Ω and 1 MΩ respectively, that the capacitor value is 1 μF, and that the square wave has a frequency of 500 Hz and a peak to peak amplitude of 10 V.

Finally, let us assume that when the generator is first turned on, its voltage jumps to +5 V, and that the capacitor voltage is zero before the generator is turned on.

What is the capacitor voltage immediately after the generator jumps to +5 V?

It is still zero volts. Therefore, the diode voltage is +5 V, which is a forward bias. Thus we can replace the diode in the circuit by a linear resistor representing its forward resistance, as in figure 49. The circuit is now the same as the simple *RC* circuit we have already discussed and so the diode voltage will decay exponentially towards zero volts.

What is the time constant of circuit?

$$\tau = CR_F$$
$$= 10^{-4} \text{ s}$$

What is the value of the diode voltage when the square wave is about to change to −5 V?

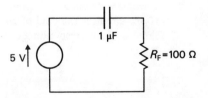

Figure 49 *A circuit representing the dc restorer when the diode is forward biased.*

A 500 Hz square wave changes from positive to negative or vice versa every millisecond. Thus the diode voltage will have decayed exponentially for a period of 10 time constants and will be effectively zero. At the same time the capacitor voltage will have exponentially increased to +5 V.

What voltages appear across the capacitor and diode immediately after the generator voltage changes to −5 V?

34

The capacitor voltage remains +5 V, so by Kirchhoff's voltage law, the diode voltage is −10 V. The diode is now reverse biased and so acts like a resistor of magnitude R_R. Figure 50 shows the circuit representing this.

Again v_D should decay to zero, but the time constant is now $\tau = R_R C = 1$ s. This is so much longer than the 1 ms during which the generator voltage remains at −5 V that by the time of the next transition to +5 V there is virtually no decay at all. At that time the diode and capacitor voltages are

$$v_D = -5 \exp(-0.001)$$
$$v_C = 10 \exp(-0.001) - 5$$

A good approximation to exp x for small values of x is given by

$$\exp(-x) \approx 1 - x$$

(This formula is explained in the section on power series in the mathematics booklet.) Thus just before the next transition of the source voltage to +5 V the circuit voltages are $v_D = -4.995$ and $v_C = 4.99$. Just after the transition, $v_C = -9.99$ and so $v_D = 5 - 4.99 = 0.01$ V.

The circuit now behaves just as after the first transition of the source to +5 V but now the diode voltage is *initially* practically zero.

From here on the cycle is repeated. The waveforms we have described are shown in figure 51.

After an initial positive spike, v_D never again becomes more than very slightly positive. Thus we say it is *clamped* to zero and the circuit is called a *clamping circuit*. The average value of v_D is just about −5 V.

clamping circuit

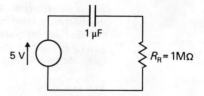

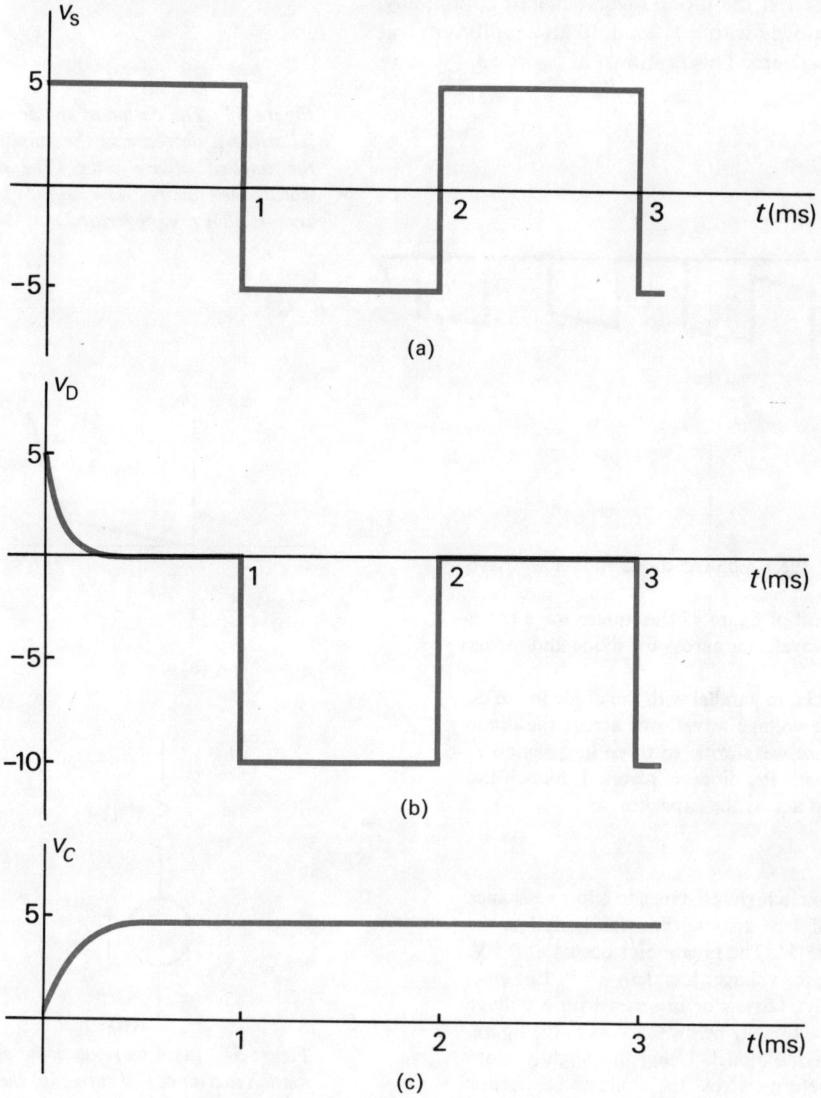

Figure 50   A circuit representing the dc restorer when the diode is reverse biased.

Figure 51   Waveforms for the dc restorer circuit. (a) source voltage; (b) diode voltage; (c) capacitor voltage.

Study comment

This is an appropriate time to do Home Experiment 2. In it you will examine the differences between the theory based upon the approximate model just described and the performance of the real circuit. You will also examine the waveforms obtained when a sine wave is applied to this circuit, and the waveforms obtained when the amplitude of the sine or square wave is changed.

You saw in the home experiment that when a 500 Hz sine wave was applied to the circuit, the diode voltage was a sine wave clamped to the same voltage as was the square wave. Can you explain this?

At any instant when the diode is reverse biased, the circuit time constant is much larger than the period of the sine wave, so the capacitor voltage remains practically constant. At any instant when the diode is forward biased, the circuit time constant is much smaller than the period of the sine wave, so the capacitor can rapidly charge up so that its voltage is very nearly the voltage of the source.

The result is that the first time the sine wave becomes positive the capacitor voltage builds up until the peak value of the sine wave is reached. When the sine wave then decreases the diode becomes reverse biased so the capacitor maintains that peak voltage. The result is an output which remains negative, except for a very brief period near each voltage peak. The waveforms are shown in figure 52.

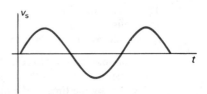
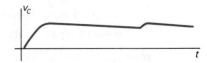
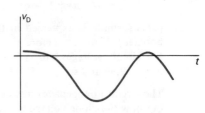

Figure 52 The output of the dc restorer when a sine wave is applied to it. (The exponential decay when the diode is reverse biased is shown exaggerated.)

Now what happened when the amplitude of either the sine or square wave was changed?

Increasing their amplitudes caused the diode to be forward biased on the first positive segment. Thereafter, the signal was immediately clamped to the new peak value. Decreasing the amplitude left the diode reverse biased continually. Thus the capacitor voltage decayed slowly with $\tau = 1$ s until an equilibrium was reached with the positive peaks just at zero. This is shown in figure 53.

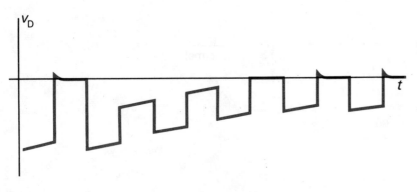

Figure 53 The output of the dc restorer following a decrease in the amplitude of the applied square wave. (The rate at which the square wave shifts towards zero is shown exaggerated.)

SAQ 6

For all of the following questions use the simplified diode model we have used up to now.
1 Suppose that in the dc restorer circuit of figure 47 the square wave has a frequency of 5 Hz. Sketch the voltage waveforms across the diode and across the capacitor.
2 Suppose that a 10 kΩ resistor is placed in parallel with the diode in the dc restorer circuit of figure 47. Sketch the voltage waveforms across the diode and across the capacitor. Compare these waveforms to those in question 1.
3 Suppose that in the dc restorer circuit the diode is reversed. Sketch the voltage waveforms across the diode and across the capacitor.

SAQ 7

To represent the fact that the change from a high resistance to a low resistance in a diode does not occur at a bias of zero assume that the diode has the V–I characteristic curve shown in figure 54. The breakpoint occurs at 0.5 V. The diode behaves like a 1 MΩ resistor for voltages less than 0.5 V. For voltages less than 0.5 V it behaves like a 100 Ω resistor in series with a voltage source whose voltage is V_D, which is fractionally below 0.5 V as in the figure. Figure 54 shows the forward biased diode model. Using this slightly more accurate model, sketch the voltage waveforms across the diode and capacitor for the dc restorer circuit of figure 47. Compare your answer with the waveforms found in Home Experiment 3.

SAQ 6

SAQ 7

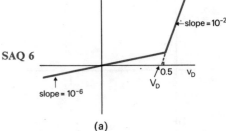
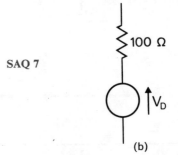

Figure 54 (a) A more accurate approximation to a diode V–I curve; (b) the circuit model representing the forward biased diode.

Section 8

6.8 RLC circuits

Now that we have covered the step response of a fairly wide range of single time constant circuits we are ready to look at the next level of complexity – circuits with two reactive elements which cannot be combined to form a single element.

Study comment

> This is an appropriate point to do Home Experiment 3. By doing it, you will see the waveforms which will now be described. In the experiment you will examine the step response of the circuit in figure 55 for various values of R.

For the larger values of R used in Home Experiment 3 the step response of the voltage across the capacitor in figure 55 appeared somewhat like that of the RC and RL circuits – an exponential rise toward its final *equilibrium value* of V_s. However, as R decreased beyond a certain value the response changed its form. The voltage across the capacitor increased beyond V_s and then oscillated about V_s. The oscillation quickly died away. As R was decreased still further the oscillation became more pronounced and died away more slowly. The curves in figure 56 are typical of the waveforms you should have seen.

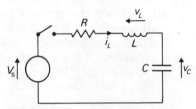

Figure 55 A series RLC circuit.

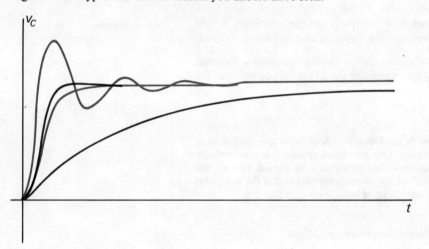

Figure 56 The step response of the capacitor voltage in a series RLC circuit for several values of R.

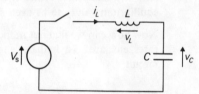

Figure 57 A series LC circuit.

When the capacitor voltage oscillates about its equilibrium value, the inductor current simultaneously oscillates about its final, or equilibrium value, which is zero. In fact, the key to understanding the oscillation is to examine the way in which the voltage across the capacitor and the current through the inductor interact. We can examine this interaction most clearly in the limiting case, in which $R = 0$. This circuit is shown in figure 57.

If the capacitor voltage is less than V_s there will be a positive voltage across the inductor equal to $V_s - v_C$. Since the rate of change of the inductor current is proportional to the inductor voltage, the inductor current will increase if v_C is less than V_s. If v_C is greater than V_s the inductor current will decrease. Similarly, if the inductor current is positive the capacitor voltage will increase because the rate of change of the capacitor voltage is proportional to the current. If the inductor current is negative, the capacitor voltage will decrease.

Let us now follow these interactions for one cycle of the oscillation, following the closing of the switch. Initially both i_L and v_C are zero. Since v_C is less than V_s, i_L starts to rise. Once i_L becomes positive it causes v_C to rise as well. Both keep rising until v_C equals V_s at which time i_L stops rising. The two waveforms are shown in figure 58. At this time they have reached point A.

So long as i_L is positive v_C keeps rising. Thus it increases past V_s and causes i_L to decrease. There is a continued rise in v_C and a continued fall in i_L until i_L reaches zero, at which time v_C stops rising. (See point B in figure 58.)

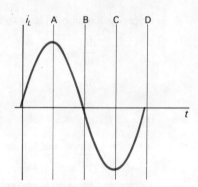

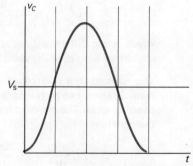

Figure 58 The capacitor voltage and inductor current in the LC circuit for one cycle of the oscillation.

That completes the first half of an oscillation cycle. At this point i_L continues to decrease since v_C is still greater than V_s. Once i_L is negative, v_L begins to fall. When v_C falls below V_s, i_L stops decreasing. (See point C in figure 58.)

In the final part of the cycle, v_C continues to fall since i_L is still negative, but i_L now increases since v_C is less than V_s. The cycle ends with both i_L and v_C at zero, and then repeats from the beginning. (See point D in figure 58.)

Notice that although the final, or equilibrium values of v_C and i_L are V_s and 0 respectively, the capacitor voltage repeatedly forces the inductor current to change *past* its final point, and vice versa. Thus we have an oscillatory situation. The question now is, what causes this oscillation to die down? As you saw in the experiment, it dies out quite rapidly when R is not zero. The key factor here is energy storage. The energy stored in the capacitor at any instant is $\frac{1}{2}Cv_C^2$, while that in the inductor is $\frac{1}{2}Li_L^2$.

Since v_C and i_L both reached their peaks at instants when the other passed its equilibrium point, the stored energy at any time was continually being passed back and forth between the capacitor and inductor. Here is where the resistor comes in. Whereas the capacitor and inductor *store* energy, the resistor dissipates it. In the limiting case where $R = 0$, no energy is dissipated and the oscillation will continue indefinitely without decreasing. However, with the resistor in the circuit, energy will be removed on each cycle. Thus the maximum swings of v_C and i_L about their equilibrium positions will get smaller on successive cycles. We say that R has damped the oscillation. When R is sufficiently large, it absorbs so much energy that there is no oscillation at all. At this point the circuit is said to be *critically damped*. Further increases in R simply cause the transition to the equilibrium values to be even slower, and the circuit is said to be overdamped.

critical damping

Now how can we find out numerically what the oscillation frequencies, damping rates, etc., are? To do that we must find and solve the differential equation of the circuit.

Study comment

> The most useful information in the following discussion is the conditions on R, L and C which determine whether the waveforms are under, over or critically damped, and the form of the expressions giving these waveforms. You will not be expected to know the details of the mathematics used to find the waveforms and in fact these details are only sketched in.

Applying Kirchhoff's laws to the circuit we get two equations

$$L\frac{di_L}{dt} = V_s - Ri_L - v_C$$

$$C\frac{dv_C}{dt} = i_L$$

Thus

$$C\frac{d^2v_C}{dt^2} = \frac{di_L}{dt}$$

Combining these gives one equation

$$V_s = LC\frac{d^2v_C}{dt^2} + RC\frac{dv_C}{dt} + v_C \tag{14}$$

which is the differential equation we need to solve. As with the RC and RL circuits we also need the values at $t = 0$ which are $v_C = 0$ and $i_L = 0$.

The solution to a differential equation of this form consists of a sum of exponential terms, $\exp pt$ (where p is a constant which must be evaluated), and a constant term. The constant term must equal V_s, in equation (14), and so leaves the equation

$$0 = LC\frac{d^2v_C}{dt^2} + RC\frac{dv_C}{dt} + v_C \tag{15}$$

to be satisfied by the exponential terms. To find the required value of p, we differentiate $\exp pt$ and substitute it into equation (15).

$$\frac{d}{dt}(\exp pt) = p \exp pt$$

$$\frac{d^2}{dt^2}(\exp pt) = p^2 \exp pt$$

So

$$0 = LCp^2 \exp pt + RCp \exp pt + \exp pt$$

Since $\exp pt$ is never zero, we can divide by it to give

$$0 = LCp^2 + RCp + 1 \qquad (16)$$

Equation (15) is a quadratic equation in p called the *characteristic equation* of the network. It is crucial to the solution since the nature of its roots determines the nature of the waveforms, as you shall see.

characteristic equation

Let us first look at the oscillatory case where $R = 0$. Then the characteristic equation becomes

$$LCp^2 + 1 = 0$$

or

$$p^2 = \frac{-1}{LC}$$

This equation has two roots

$$p_1 = \left(\frac{-1}{LC}\right)^{1/2}$$

$$= \frac{j}{(LC)^{1/2}},$$

$$p_2 = -\left(\frac{-1}{LC}\right)^{1/2}$$

$$= \frac{-j}{(LC)^{1/2}}$$

Since there are two roots the form of the solution is

$$v_C = A_1 \exp p_1 t + A_2 \exp p_2 t + V_s \qquad (17)$$

We find A_1 and A_2 by using conditions at $t = 0$.

$$A_1 = A_2$$

$$= -\frac{V_s}{2}$$

The solution is thus

$$v_C = \frac{V_s}{2}\left\{-\exp\frac{jt}{(LC)^{1/2}} - \exp\frac{-jt}{(LC)^{1/2}}\right\} + V_s$$

$$= V_s\left\{1 - \cos\frac{t}{(LC)^{1/2}}\right\}$$

(The cosine function was obtained by combining the two complex exponentials. Read section 2.7 of the mathematics booklet if you want an explanation of this step.)

We let $\omega_0 = 1/(LC)^{1/2}$, a number called the *natural frequency** of the circuit. Thus

natural frequency

$$v_C = V_s(1 - \cos \omega_0 t)$$

The response is the sum of a constant, V_s (the equilibrium value of v_C) and a sinusoid† with an amplitude of V_s and a frequency of ω_0.

---

* As defined, this is an angular frequency, in rad s^{-1}. The natural frequency in hertz is $f_0 = 2\pi/(LC)^{1/2}$.

† The term 'sinusoid' is used to describe cosine as well as sine functions. A cosine function is simply a sine function shifted along the axis.

If R is not zero, some energy will be absorbed and the sinusoid will be damped exponentially. The solution is still equation (17) but the constants p_1, p_2, A_1 and A_2 are different. Let us introduce one more new constant called the *damping factor*, which is given by

damping factor

$$K = R/2L$$

If the characteristic equation is rewritten using K and ω_0 it becomes

$$p^2 + 2Kp + \omega_0^2 = 0$$

Its roots are

$$p_1 = -K + (K^2 - \omega_0^2)^{1/2}$$
$$p_2 = -K - (K^2 - \omega_0^2)^{1/2} \qquad (18)$$

These roots may be real or complex, depending upon the values of K and ω_0. If they are complex, then when they are substituted into equation (17), the imaginary parts will simplify into a sinusoidal term, just as for $R = 0$, i.e. there will be an oscillation. If the roots are real, equation (17) will simply be the sum of two exponentials and so there will not be any oscillation. From equation (18) the roots will be complex if $K^2 < \omega_0^2$, or in terms of R, L and C, $R < 2(L/C)^{1/2}$.

We have said that the solution will be a damped sine wave for 'small R'. This equation tells exactly how small it must be.

For $K^2 > \omega_0^2$ (or $R > 2(L/C)^{1/2}$) the roots will be real and negative. For $K = \omega_0$ ($R = 2(L/C)^{1/2}$) the borderline case, there will be one real, negative root, i.e. $p_1 = p_2 = -K$.

The solutions for these three cases are now listed.

Case 1

$R < 2(L/C)^{1/2}$ or $K^2 < \omega_0^2$. This case is usually called *underdamped* (in contrast to the next case)

underdamped step response

$$v_C = V_s \left\{ 1 - \frac{\omega_0}{\omega_d} (\exp -Kt) \cos(\omega_d t + \phi) \right\}$$

where $\omega_d = (\omega_0^2 - K^2)^{1/2}$ and where $\phi = \tan^{-1} \frac{K}{\omega_d}$.

The waveform oscillates because of the cosine term. The oscillation frequency is ω_d. Notice that since ω_d is less than ω_0 the presence of the resistor not only damps the amplitude of the oscillation but also lowers its frequency. The oscillation is exponentially damped by $\exp -Kt$. As the exponential gets small all that is left is the final or equilibrium value of the capacitor voltage, which is V_s. The waveform is shown in figure 59. It should be clear why K is called the damping factor.

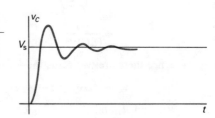

Figure 59 *The step response of the capacitor voltage for an underdamped RLC circuit.*

Case 2

$R = 2(L/C)^{1/2}$ or $K = \omega_0$. This case is called *critically damped* as the damping is just sufficient to eliminate the oscillation.

critically damped step response

$$v_C = V_s\{1 - (1 + Kt) \exp -Kt\}$$

Without the multiplying factor $(1 + Kt)$ this expression would be identical to the step response of the capacitor voltage in the RC circuit. This multiplying factor affects the waveform most strongly when t is small as you can see by comparing its graph in figure 60 with figure 31. As t gets large the exponential term approaches zero much more rapidly than $(1 + Kt)$ gets large, so the voltage approaches V_s.

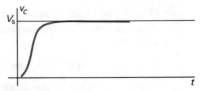

Figure 60 *The step response of the capacitor voltage for a critically damped RLC circuit.*

Case 3

$R > 2(L/C)^{1/2}$ or $K > \omega_0$. This is the *overdamped case*. The exponential rise is slower than for the critically damped circuit.

overdamped step response

$$v_C = V_s \left(1 - \frac{p_2}{p_2 - p_1} \exp p_1 t - \frac{p_1}{p_1 - p_2} \exp p_2 t \right)$$

where $p_1 = -K + (K^2 - \omega_0^2)^{1/2}$ and $p_2 = -K - (K^2 - \omega_0^2)^{1/2}$.

This expression is the sum of two exponentials and a constant. Both p_1 and p_2 are negative* so that, again, the exponentials decay as t increases, leaving just the constant V_s as the final value of v_C. As in the critically damped case the waveform (see figure 61) has more or less the same appearance as the single time constant exponential rise in the RC circuit.

There are some special points to notice. Suppose you were trying to choose R, L and C so that the output was approximately like the step input. How would you obtain the minimum *rise time* (usually defined as the time required for the output to change from 0.1 to 0.9 of its final value)? Certainly critical damping is better than overdamping in this respect, but if a slight amount of oscillation can be tolerated in a given application, slight underdamping will give an even faster rise time. With a slightly underdamped circuit so that the oscillation is about 10% of the total step, the rise time is cut by about one half its value for a critically damped circuit. (You can see this effect in figure 56.)

As with the single time constant circuits, there is also a large class of circuits which behave similarly to the series RLC circuit we just examined. However, there are no simple shortcuts which enable them to be analysed by direct analogy as with the single time constant circuits. (An exception is the parallel RLC circuit examined in SAQ 8.) In general, you must set up the differential equation and find the characteristic equation. Once this is done, the roots of the characteristic equation tell whether the circuit is over, under, or critically damped, but you must still use the initial conditions to find out what all the constants are.

rise time

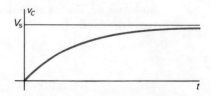

Figure 61 *The step response of the capacitor voltage for an overdamped RLC circuit.*

SAQ 8

The switch in the parallel RLC circuit below opens at $t = 0$. Using the principle of duality find (a) the conditions which determine when the step response of the inductor current is under, over and critically damped, (b) the inductor current waveform for the underdamped case.

SAQ 8

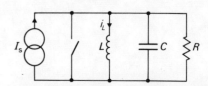

Figure 62 *A parallel RLC circuit.*

Exercises:

1 Find an expression for the step response of the inductor current in the series RLC circuit for the overdamped case. (Hint: Use the result obtained for the capacitor voltage.)

2 For a series RLC circuit, $L = 100$ mH, $C = 0.1$ μF, $R = 100$ Ω. What is (a) the natural frequency (b) the damping constant and (c) ω_d? After how many cycles will the oscillation have damped to 0.1 of its original amplitude? (Hint: $\exp(-2.3) \simeq 0.1$.)

1 Using $i_L = C\, dv_C/dt$

$$i_L = V_s C \frac{p_1 p_2}{p_1 - p_2} (\exp p_1 t - \exp p_2 t)$$

2 $\omega_0 = 10^4$, $K = 500$, $\omega_d = 0.999 \times 10^4$. After 8 cycles the oscillation will have an amplitude less than one tenth its original amplitude.

6.8.1 Summary of section 6.8

The step response of a series RLC circuit can take three distinct forms depending upon the relative values of R, L and C.

1 Underdamped: when $R < 2(L/C)^{1/2}$ the response is an exponentially damped sinusoid.

$$v_C = V_s \left\{ 1 - \frac{\omega_0}{\omega_d} (\exp -Kt) \cos(\omega_d t + \phi) \right\}$$

In the limiting case where $R = 0$ (i.e. an LC circuit) the sinusoid is not damped at all. That is $\omega_d = \omega_0$ and K and ϕ are zero so

$$v_C = V_s(1 - \cos \omega_0 t)$$

* *They must be negative because $K^2 - \omega_0^2 < K^2$ so $(K^2 - \omega_0^2)^{1/2} < K$.*

2. Critically damped: when $R = 2(L/C)^{1/2}$ there is no oscillation at all.

$$v_C = V_s\{1 - (1 + Kt)\exp -Kt\}$$

3. Overdamped: when $R > 2(L/C)^{1/2}$ the rise in capacitor voltage is slower than when critically damped.

$$v_C = V_s\left(1 - \frac{p_2}{p_2 - p_1}\exp p_1 t - \frac{p_1}{p_1 - p_2}\exp p_2 t\right)$$

The conditions determining the nature of the step response are found from the roots of the characteristic equation. When there are two complex roots the circuit is underdamped. When there are two real roots the circuit is overdamped. For the borderline case, when there is one real root the circuit is critically damped.

Section 9

6.9 Circuit transient or natural response

Throughout this unit we have been looking at the response of a variety of electronic circuits to a particular signal – a step in voltage or current. That is, we were discussing 'signal processing'. Now let us look at the same topic from a different point of view.

Before the step was applied, we assumed that the circuit was in equilibrium – that is, that none of the voltages or currents were changing.

After the step was applied each circuit responded to this signal in its own way but eventually settled down to a new equilibrium, where again no voltages or currents were changing. Thus the application of a step in voltage or current can be thought of as a change in the equilibrium conditions. The waveforms we examined showed how each circuit adjusted itself to these new equilibrium conditions. For this reason, the step response is also called the *natural response* of the system.

natural response

Networks containing only resistive elements respond to a change in equilibrium conditions instantaneously. Networks containing reactive elements – capacitors and inductors – do not. A capacitor or inductor stores energy. When the equilibrium conditions are changed this energy redistributes itself in accordance with the new conditions. It is this shift of energy that takes time. Resistive elements do not store energy and so can respond instantly.

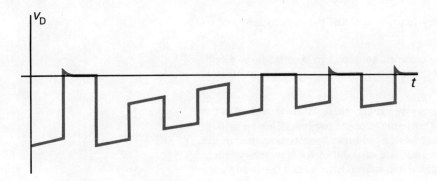

Figure 63

Now suppose a circuit contains a voltage or current source with a periodic signal, like a sine or square wave. The response of the circuit will change when that signal is turned on or off or is changed in amplitude. This change in response is also characterized by the natural response of the circuit. For example, in figure 63 I have repeated the waveform of figure 53 showing the response of a dc restorer to a decrease in the amplitude of a square wave. The transition from the large clamped square wave to the small clamped square wave consists of a square wave plus a decaying exponential. The latter is the natural response of the dc restorer circuit when the diode is reverse biased.

In contrast to the natural response of a circuit, its response to a varying applied signal itself is called its *forced response*. In a practical situation where the varying signal is turned on at some time the total response of the network will combine the forced response to that signal with the transient response to account for the transition from off to on.

forced response

For any linear circuit it is quite straightforward to separate the total response to some signal into the forced and natural responses, in terms of the solutions to its differential equation.

For example, consider the following differential equation

$$a\frac{dx}{dt} + bx = f(t)$$

The term $f(t)$ is the forcing function. (It would often be a sine or square wave.)

The complete solution of this equation contains two parts called the *particular solution* and the *complementary function* respectively. The particular solution x_p is some function which satisfies the differential equation alone, but not the conditions at $t = 0$.

$$a\frac{dx_p}{dt} + bx_p = f(t) \tag{19}$$

It represents the forced response of the circuit.

The complementary function x_c is some function which satisfies the differential equation with $f(t)$ set to zero.

$$a\frac{dx_c}{dx} + bx_c = 0 \tag{20}$$

It represents the natural response of the circuit and is the same for any forcing function.

The total solution to the differential equation x_t is the sum of the particular solution and the complementary function multiplied by a constant A, which is determined by the condition at $t = 0$.

$$x_t = x_p + Ax_c$$

To see how this works add equations (19) and (20)

$$a\left(\frac{dx_p}{dt} + \frac{d(Ax_c)}{dt}\right) + B(x_p + Ax_c) = f(t)$$

$$\left(a\frac{dx_p}{dt} + bx_p\right) + A\left(a\frac{dx_c}{dt} + bx_c\right) = f(t)$$

$$f(t) + A \times 0 = f(t)$$

In many circuits the complementary function eventually decays to zero and so represents a transient waveform.

Thus by containing two terms, the complete solution can account for not only the response to the forcing signal but the initial conditions and a smooth transition between the two.

There are many applications, with varying signals, where we are interested in the forced response to these signals and not in the transient response. Thus we usually make the assumption that the signal was turned on at some remote time in the past so that all the transient waveforms have died out by the time the system is observed. The system then settles down into what is often called the steady state response. This is precisely what we shall do in the next course unit. We shall apply sine waves to a variety of circuits and shall discuss not the transient part of the waveforms which result but just the *sinusoidal* steady state.

Section 10

6.10 Summary of unit

This is the first unit which has concentrated on electronic components connected together to form circuits, rather than on the operation of the components.

The first concepts discussed were the fundamental relations between currents and voltages in connected elements, Kirchhoff's laws.

Kirchhoff's current law: The sum of the currents entering any node is zero at all times.

Kirchhoff's voltage law: The sum of the voltages around any closed path is zero at all times.

These two laws apply equally to constant and varying voltages and currents, and to circuits containing any combination of components.

With Kirchhoff's laws, and the behaviour of components from the first five course units to build upon, we began the main topic of the unit, *signal processing*, or the way in which different circuits respond to sources with various waveforms.

We first looked at resistive circuits, which are circuits where all currents and voltages depend only upon the instantaneous values of the source currents and voltages and not upon the way in which the sources are varying. Once you have found an expression for any voltage or current in a resistive circuit with constant sources you can find the corresponding expression for time varying sources simply by replacing the constant with the varying signal in the original expression.

The most important topic of the unit was the response of single time constant circuits to a voltage or current step. Any circuit which can be reduced to a single capacitor or inductor plus resistors and sources is a single time constant circuit.

If you know the initial and final values of the capacitor voltage or inductor current and the time constant, the response to a step is given by the formula

$$y = (y_i - y_f) \exp \frac{-t}{\tau} + y_f$$

For the capacitor voltage in the simple RC circuit

$$v_C = V_s \left(1 - \exp \frac{-t}{RC}\right)$$

For the inductor current in the simple RL circuit

$$i_L = I_s \left(1 - \exp \frac{-tR}{L}\right)$$

The response of a nonlinear circuit, the dc restorer, can be explained in terms of two single time constant circuits by using a simplified model of the diode. In the first circuit the diode is replaced by a resistor representing its forward resistance. In the second the diode is replaced by a resistor representing its reverse resistance.

The step response of an RLC circuit takes three forms, depending upon the values of R, L and C.

If $R < 2(L/C)^{1/2}$ the capacitor voltage and inductor current oscillate before settling down to their final values. The circuit is underdamped.

If $R > 2(L/C)^{1/2}$ the capacitor voltage rises towards its final value without oscillating. The circuit is overdamped.

With $R = 2(L/C)^{1/2}$, the borderline case, the circuit is critically damped.

The response of a circuit to a voltage or current step can be thought of as the natural response of the circuit to a change in equilibrium conditions. The total response of a linear circuit to a time varying signal consists of two parts, the forced response (to the signal itself), and the natural response (to account for the transition from the initial conditions to the steady state).

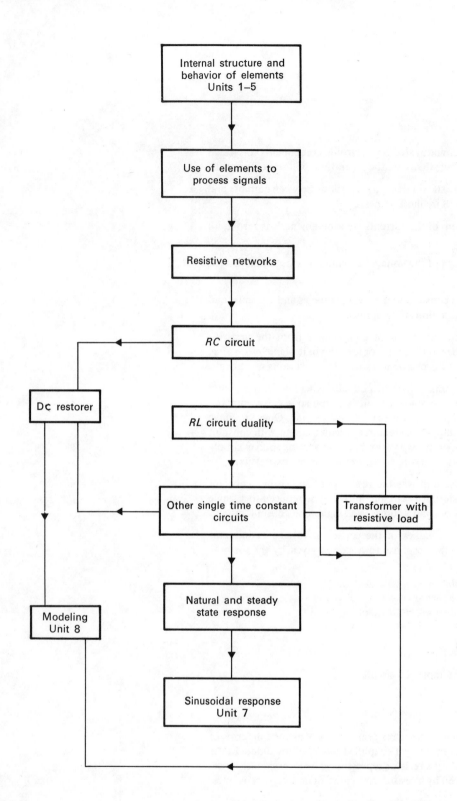

Self-assessment questions

Question 9

In the circuit of figure 64, N_1, N_2, N_3 and N_4 are parts of a circuit. The wires shown lead from components in one part to components in another. Find i_1, i_2 and i_3.

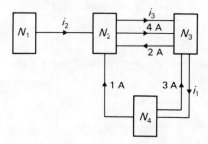

Figure 64

Question 10

In the circuit of figure 65 find the current through the 1 Ω resistor and the voltage across the 12 Ω resistor.

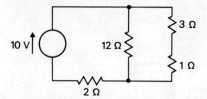

Figure 65

Question 11

In the circuit of figure 66 the switch is initially at position 2. It changes to position 1 at $t = 0$ and then returns to position 2 at $t = 1$ ms. Calculate and sketch the waveform which appears across the capacitor. (Assume that $v_C = 0$ at $t = 0$.)

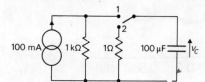

Figure 66

Question 12

In the circuit of figure 67 the voltage source puts out a square wave which alternates between 0 V and 4 V, with a frequency of 2 kHz. It is desired to apply a square wave to the amplifier which has a dc level of zero and so the coupling capacitor C is used. If the amplifier behaves like a 10 kΩ resistor from its input terminals, what is the minimum value of C required so that the waveform at v_{in} does not decay by more than 1% during each half cycle of the square wave input?

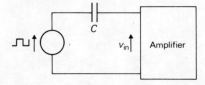

Figure 67

Self-assessment answers and comments

Question 1

(a) $v = iR$ (b) $v = L\,di/dt$ (c) $i = C\,dv/dt$

Question 2

The total resistance R is given by

(a) $R = R_1 + R_2$

(b) $\dfrac{1}{R} = \dfrac{1}{R_1} + \dfrac{1}{R_2}$

Question 3

The voltage across the 3 Ω resistor is $3 \times 6 = 18$ V. Thus

$$V_1 = 18 + 12$$
$$= 30\ \text{V}$$

The current through the 4 Ω resistor is $12/4 = 3$ A. Thus

$$I_1 = 3 - 6$$
$$= -3\ \text{A}$$

Question 4

1. By comparing values of $v_{in(1)}$ and $v_{out(1)}$ for corresponding points in time we can obtain the input–output curve shown in figure 68. Using it, the response to $v_{in(2)}$ is as shown in figure 69.
2. After $t = 2$, $v_{in(2)}$ repeats values it has previously taken. For example, it equals 0 V at $t = 1$ and $t = 3$. However, $v_{out(2)}$ does not repeat at corresponding instants. At $t = 1$, $v_{out(2)} = 1.5$ V but at $t = 3$, $v_{out(2)} = 2$ V. Thus there is no single input–output relation as required for a resistive network, so $v_{out(2)}$ could not be a response to $v_{in(1)}$.

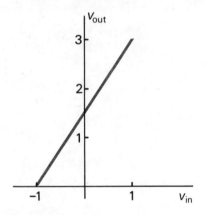

Figure 68

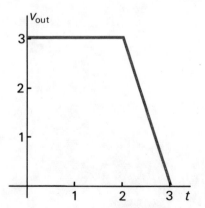

Figure 69

Question 5

(a) The initial voltage of the capacitor is 3 V, the final voltage is -3 V and the time constant is 1 s.

$$v_C = -3 + 6\exp -t$$

(b) Since the time constant is 1 s, by $t = 5$ s the capacitor voltage will be almost equal to its final value of -3 V. (Remember Table 1!) This becomes the new initial value. The new final value is the same as the original initial value of $+3$ V. The new time constant is still 1 s.

$$v_C = +3 - 6\exp(-t + 5)$$

for $t > 5$

(c) Continuing as above gives the waveform of figure 70.

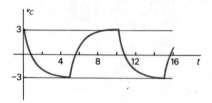

Figure 70

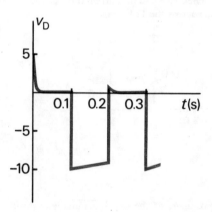

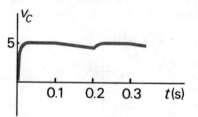

Figure 71

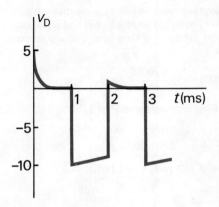
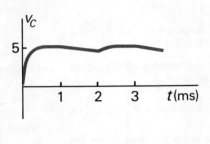

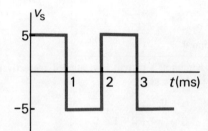

Figure 72

Question 6

1 Figure 71 shows the answer. Because the square wave frequency is so much lower the capacitor voltage decays by about 10% when the diode is reverse biased.
2 Figure 72 shows the answer. Because of the 10 kΩ resistor the time constant when the diode is reverse biased is $t = 1\,\mu\text{F} \times 10\,\text{k}\Omega = 0.01$ s. As in question 1, there will be a 10% decay in capacitor voltage when the diode is reverse biased.
3 Figure 73 shows the answer. The diode will be forward biased the first time the square wave becomes negative. The capacitor will charge up to that negative value and will hold it when the square wave subsequently becomes positive. The result is a diode voltage clamped to zero with a positive dc level.

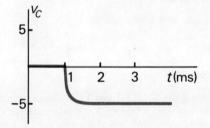

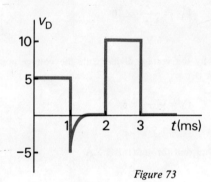

Question 7

Immediately following the initial transition of the generator output to $+5$ V the diode is forward biased and the circuit appears as in figure 74. The capacitor voltage charges exponentially towards $5 - V_D$, which is just over 4.5 V, with a time constant $\tau = R_F C = 0.1$ ms. Once it passes 4.5 V the diode acts like a 1 MΩ resistor. The capacitor voltage continues rising exponentially towards 5 V but with a time constant $t = R_R C = 1$ s. Thus it remains virtually constant at 4.5 V. From this point onwards the argument proceeds as for the dc restorer with the simpler diode model. The diode voltage waveform is now clamped to 0.5 V, as shown in figure 74(b).

Figure 73

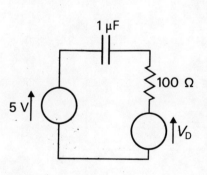

Figure 74 (a)

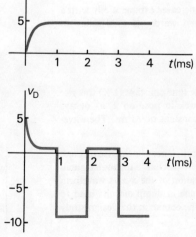

Figure 74 (b)

Question 8

Since the parallel RLC circuit is the dual of the series RLC circuit the answers are found by replacing L by C and vice versa, R by $1/R$ and vice versa, etc. Thus for the underdamped case, the condition is $1/R < 2(C/L)^{1/2}$ or $R > \frac{1}{2}(L/C)^{1/2}$.

Similarly, the condition for critical damping is $R = \frac{1}{2}(L/C)^{1/2}$ and for an overdamped response $R < \frac{1}{2}(L/C)^{1/2}$.

Notice that a *large* resistance is required for oscillation in the parallel RLC circuit. The current waveform is

$$i_L = I_s \left\{ 1 - \frac{\omega_0}{\omega_d} (\exp -Kt) \cos(\omega_d t + \phi) \right\}$$

where ω_0, ω_d and ϕ are defined as for the series circuit but $K = 1/2RC$.

Question 9

Using Kirchhoff's current law, applied respectively to N_1, N_2 and then N_4 we get

$$i_2 = 0$$
$$i_3 = -4 + 2 + 1 + 0$$
$$= -1$$
$$i_1 = 1 + 3$$
$$= 4$$

Question 10

Combining the 12 Ω, 3 Ω and 1 Ω resistors gives an equivalent resistance of 3 Ω. By the voltage divider rule, the voltage across this combination (and thus also across the 12 Ω resistor) is

$$V_{12} = \frac{3}{3+2} \times 10$$
$$= 6 \text{ V}$$

Again by the voltage divider rule the voltage across the 1 Ω resistor is

$$V_1 = \frac{1}{1+3} \times 6$$
$$= 1.5 \text{ V}$$

so the current through it is 1.5 A.

Question 11

With the switch in position 1, v_C increases exponentially with a time constant of 0.1 s from zero towards its final value of 100 mA × 1 kΩ = 100 V.
At $t = 1$ ms it has reached

$$100(1 - \exp 0.01) \simeq 1 \text{ V}$$

Since 1 ms is a small fraction of the time constant (1%) this rise is quite linear. With the switch back in position 2, v_C decays exponentially to zero with a time constant of 0.1 ms. The waveform is shown in figure 75.

The principle illustrated by this circuit, a near linear rise through a large resistance followed by a rapid discharge through a small resistance is the basis of the generation of the x-axis waveform of an oscilloscope. You will examine in depth circuits used to generate such waveforms later in the course in the design study on time base generators.

Question 12

For v_{in} to approximate a square wave the time constant must be long compared with 0.25 ms, the half period of the square wave. The capacitor voltage will then remain nearly constant. For a 1% change

$$0.99 = \exp \frac{0.00025}{RC}$$

Using $\exp(-x) \simeq 1 - x$ for small x

$$0.01 = \frac{0.00025}{RC}$$

With $R = 10$ K this gives $C = 2.5$ μF.

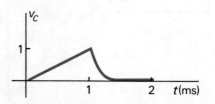

Figure 75

Unit 7

Signal Processing II
Sinusoidal Response

CONTENTS

	Aims and objectives	4
7.1	Introduction	5
7.2	**Linearity**	6
7.2.1	Linear components	6
7.2.2	Properties of linear networks	7
7.2.3	Summary of section 7.2	8
7.3	**Sinusoidal response of linear components**	11
7.3.1	Sinusoidal signals	11
7.3.2	Sinusoids in electronic components and circuits	12
7.4	**Using complex numbers to handle sinusoids**	15
7.4.1	Phasor notation	15
7.4.2	Exercises on phasor notation	17
7.4.3	Complex impedance	17
7.4.4	The impedances of combinations of components	18
7.4.5	A worked example: an RL circuit	19
7.4.6	Exercises on impedances and phasors	20
7.4.7	Summary of sections 7.3 and 7.4	23
7.5	**Frequency characteristics of circuits**	25
7.5.1	Describing frequency characteristics	25
7.5.2	Frequency response of RL and RC circuits	27
7.5.3	Frequency response of an RLC circuit	31
7.6	**Fourier series**	37
7.7	**Summary of Unit 7**	42
	Self-assessment questions	44
	Self-assessment answers and comments	44
	Appendix 1 Obtaining Fourier coefficients	47

Aims

The aims of this unit are:

1 To establish the concept of a linear circuit and to show why its properties of superposition and homogeneity are such powerful analytical tools.

2 To explain why linear electrical networks are characterized by their response to sinusoids.

3 To explain the use of the mathematical techniques used to find the sinusoidal response: phasors and complex impedances.

4 To show how phasors and complex impedances can be used to analyse and design *RL*, *RC*, *RLC* high pass, low pass and band pass circuits.

Objectives

When you have completed this unit you should be able to:

1 Define, or recognize adequate definitions of, or distinguish between true and false statements concerning each of the following terms, concepts or principles:

admittance	linear circuit
amplitude	linearity
angular frequency	low pass circuit
band pass circuit	nonlinear circuit
band width	phase angle
cut-off frequency	phase lag
decibel	phase lead
Fourier series	phasor
fundamental frequency	Q (quality factor)
half power frequency	resonance
harmonic	resonant frequency
high pass circuit	sinusoid
homogeneity	superposition
impedance	

2 State the properties of linear circuits and sinusoids which allow you to determine the response of a linear circuit to a periodic input signal.

3 Find the phasor for any sinusoidal function.

4 Calculate the complex impedances of combinations of resistors, capacitors and inductors.

5 Find the steady state voltages and currents in simple network configurations with applied sinusoids.

6 Plot the frequency response curves of *RL*, *RC* and *RLC* circuits.

7 Design simple high, low, and band pass circuits using *RC*, *RL* and *RLC* circuits.

Section 1

7.1 Introduction

In Unit 6 you began the study of signal processing. You saw how an important but rather restricted class of circuits processed voltage and current steps. In this unit we will look at how a considerably wider but still restricted class of circuits, linear circuits, process sinusoids. Since our goal is to develop methods to enable us to understand how virtually any circuit processes any signal, you might begin to wonder whether we won't run out of course units before we run out of types of circuits and signals.

In fact the techniques to be introduced in this unit are extremely powerful. A knowledge of the response of a linear network to any sinusoid can be used to predict its response to any other signal you will meet. Moreover, as promised in Unit 6, these techniques will not require the solution of a differential equation. That step is eliminated. In contrast, no similar techniques are available for *nonlinear* circuits and the response of a nonlinear circuit to a given signal usually must be found separately. For this reason, one often tries to use a linear circuit to approximate a nonlinear circuit (you saw one way of doing this in Unit 6 when we discussed the dc restorer), even if the linear model is a very poor approximation. It is assumed that a solution which is inaccurate is better than no solution at all, although that assumption doesn't always work!

To find the sinusoidal response of a linear network we will introduce two concepts: phasors to characterize sinusoids and complex impedance to characterize the relationships between currents and voltages of components or circuits. Your main objectives in this unit should be to understand what these two concepts mean and, by doing the exercises and SAQs, to become familiar with their use. To do this you will need to know most of the material in the sections of *Mathematics for Electronics* on sine waves and on complex numbers. Be sure you understand these sections before reading sections 7.3 to 7.6 in this unit.

Incidentally, for the numerical calculations in this unit you will save yourself a lot of time by using a slide rule rather than tables of logarithms and trigonometric functions. You are better off spending your time learning concepts than on doing arithmetic. When you are asked to draw graphs just calculate a few points and draw a rough curve. When you obtain answers in terms of π or square roots, don't bother to evaluate them numerically – $5/\pi$ or $4\sqrt{5}$ are acceptable answers and needn't be calculated. Also, you may find it simpler to do calculations on angles in degrees rather than in radians which I tend to use.

The purpose of the final section on Fourier series is to show how sinusoids of different frequencies can be combined to form other signals. It is the basis of the idea that the response of a linear network to sinusoids can be used to find its response to other signals. It is less important than the earlier sections and can be omitted if you are behind.

The home experiments illustrate the ideas in several sections but can also be omitted if you need to spend extra time working the exercises.

Section 2

7.2 Linearity

The main results in this course unit only apply to a class of circuits called *linear circuits*. Let us now find out just what a linear circuit is and what properties make it so useful.

linear circuits

7.2.1 Linear components

In Unit 6 we contrasted a resistor, whose *V–I* characteristic curve is the straight line $V = IR$, with nonlinear devices such as a diode, whose *V–I* characteristic curve is not a straight line. Since its characteristic curve is a straight line, a resistor is a *linear device*.

linear device

The slope of the resistor's characteristic curve is R, its resistance. The slope of the diode characteristic at any one point is the ratio of an incremental change in voltage δV to an incremental change in current δI and so represents an *incremental resistance**. This incremental resistance varies from point to point along the curve.

incremental resistance

To determine whether or not a capacitor is a linear device we also examine its characteristic curve. In contrast to that of a resistor, a capacitor's characteristic curve is a plot of charge against voltage.

All capacitors discussed so far in this course have been linear because their charge–voltage characteristic curves are straight lines, given by the equation $Q = CV$.

However, there are certain semiconductor diodes which are used as capacitors and which have nonlinear charge–voltage characteristic curves, the slopes of which represent an incremental capacitance which varies with the voltage across the device.

The characteristic curve of an inductor which determines its linearity is a plot of flux against current. A linear inductor has a flux–current characteristic curve which is the straight line $\Phi = LI/N$, where N is the number of turns. This equation represents an inductor with an air core.

> What is the shape of the flux–current characteristic of an inductor with a core made of a magnetic material?

As you learnt in Unit 4, the *B–H* curve is a hysteresis curve. Since the flux is proportional to *B* and current is proportional to *H*, the flux–current curve is also a hysteresis curve. Thus any inductor with a core of magnetic material is nonlinear. However, it is very difficult to form an adequate model of a magnetic hysteresis loop, but if the hysteresis loop is relatively narrow, and if the current through the inductor is kept small enough so that the flux does not reach the saturation region of the hysteresis curve, then the curve can be approximated by a straight line with reasonable success.

> Are transistors linear devices?

No they are not. Since transistors are resistive devices, their *V–I* characteristics determine whether or not they are linear and these consist of families of curves which are not straight lines. Certain applications of transistors, such as

---

* *The slope represents a resistance if the voltage is plotted along the vertical axis and current along the horizontal axis. If the axes are reversed, as is often the case, the slope represents an incremental conductance.*

switching or logic circuits, depend for their operation on the nonlinearity of transistor characteristics. However, transistors are also widely used in circuits intended to be linear. Such circuits can be made to behave linearly by arranging the components to at least partially compensate for the nonlinearity of transistors or by suitably restricting the transistor currents and voltages so that over a limited portion of the characteristic they behave nearly linearly. Details of the latter technique are discussed in Unit 8.

7.2.2 Properties of linear networks

Any circuit made up of linear components or of linear approximations to nonlinear components is a linear circuit. It will have two very useful properties due to its linearity.

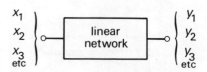

Figure 1 *A linear network with several possible inputs and their corresponding outputs.*

I have schematically indicated a linear network in figure 1, with several possible inputs, x_1, x_2, x_3, etc. (They may either be voltages or currents, depending upon what the network actually is.) For each of them, the network responds with a corresponding output, y_1, y_2, y_3, etc.

The first useful property is the *principle of homogeneity*.

principle of homogeneity

If the input signal of a linear network is the product of a constant and some other signal then the output will be the response to that other signal multiplied by the same constant.

If the input to the system in figure 1 is ax_1, the output will be ay_1.

The second property is the *principle of superposition*.

principle of superposition

If the input of a linear network is the sum of two signals, the output will be the sum of the two output signals found if each input signal were applied alone.

If the input to the system in figure 1 is $x_1 + x_2$, the output will be $y_1 + y_2$. It is quite easy to show that the principle of superposition also applies to sums of more than two inputs.

Now let us examine the implications of these properties by looking at some examples. We will start with the nonlinear circuit containing a diode and voltage source as in figure 2(a). The diode characteristic is shown in figure 2(b). You can see that doubling the source voltage from V_1 to V_2 causes the current to increase from I_1 to I_2, which is far more than double. Thus the principle of homogeneity does not apply to this nonlinear circuit.

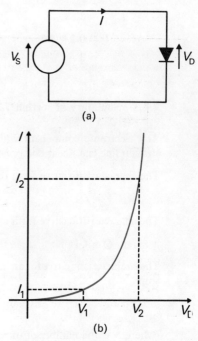

Figure 2 *A nonlinear circuit showing that the principle of homogeneity doesn't apply.*

Try a similar calculation using a resistor instead of the diode. You will see why the principles of superposition and homogeneity only work for a characteristic which is a straight line.

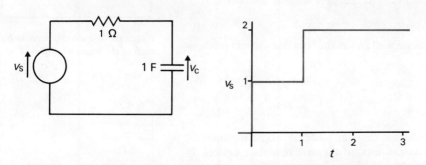

Figure 3 *An RC circuit with a double step voltage applied to it.*

Now let us examine the RC circuit in figure 3(a) to which we have applied the double step of figure 3(b). We can break the double step into two single steps (figure 4) and find the response to each step as in Unit 6. The total response is found by adding these two, as shown in figure 4, because of the principle of superposition.

The principle of superposition applies not only to signals which are the sum of two other signals, but to two or more signals from different sources.

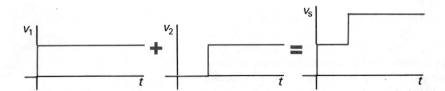

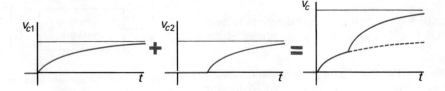

Figure 4 By the principle of superposition, the response to the sum of two signals can be found by adding the responses to each separate signal.

Let us see how it can help us in a circuit containing two sources. In figure 5 the voltage V_1 and current I_1 can be found by adding the voltages and currents caused by each source separately. If the current source is set to zero the circuit appears as in figure 6(a). The voltage across each 10 Ω resistor will be 5 V. If the voltage source is set to zero the two resistors will be in parallel as shown in figure 6(b). The current through each will be 2.5 A. Thus the voltage across them will be 25 V. Therefore

$$V_1 = 5 + 25$$
$$= 30 \text{ V}$$
$$I_1 = 0.5 + 2.5$$
$$= 3 \text{ A}$$

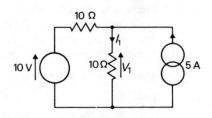

Figure 5 A circuit containing two sources.

7.2.3 Summary of section 7.2

Any electronic component with a characteristic curve (or curves) which is a straight line is a linear component.

The characteristic curve for a resistor is its voltage–current relationship

$$V = IR$$

The characteristic curve for a capacitor is its charge–voltage relationship

$$Q = CV$$

The characteristic curve for an inductor with an air core is its flux–current relationship

$$\Phi = \frac{LI}{N}$$

(where N is the number of turns).

The characteristic curves of inductors with cores made of magnetic materials, diodes, transistors, etc. are not straight lines, so they are nonlinear devices. In many applications approximate models which are linear are used to represent nonlinear devices.

A circuit representation containing linear components and models of nonlinear components which are approximately linear, is a linear circuit. There are two important properties of linear circuits.

The principle of homogeneity

If the input signal of a linear network is the product of a constant and some other signal, then the output will be the response to that other signal multiplied by the same constant.

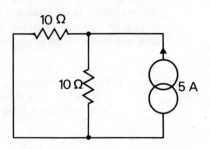

Figure 6 The circuits formed by setting each of the sources in figure 5 to zero.

The principle of superposition

If the input of a linear network is the sum of two signals, the output will be the sum of the two output signals found if each input signal were applied separately.

These two principles enable us to easily find the response of a linear circuit to a signal composed of sums and multiples of other signals for which we already know the response. It also simplified the problem of finding the response to a linear circuit with several sources. The response to each can be found separately and the results added.

Exercises

For each of the circuits in figure 7 find the voltage V.

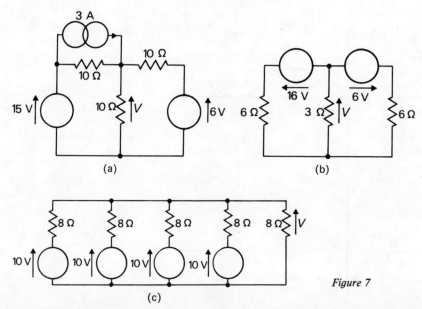

Figure 7

(a) First find the voltage across the resistor where the voltage V is indicated due to each source separately. To find the voltage due to the current source set both voltage sources to zero (see figure 8(a)). The three resistors are then in parallel. Since they are all 10 Ω, one third of the current, (i.e. 1 A.) flows through each. The voltage due to the current source is

$$1 \text{ A} \times 10 \text{ Ω} = 10 \text{ V}.$$

To find the voltage due to the 15 V source set the other two sources to zero (see figure 8(b)). The two 10 Ω resistors in parallel are equivalent to a single 5 Ω resistor. They are in series with a 10 Ω resistor. By the voltage divider rule, the voltage V due to the 15 V source is

$$\frac{5}{5+10} \times 15 = 5 \text{ V}$$

To find the voltage due to the 6 V source set the other two sources to zero (see figure 8(c)). By the same argument as for the 15 V source, the voltage V due to the 6 V source is

$$\frac{5}{5+10} \times 6 = 2 \text{ V}$$

By the principle of superposition the total voltage V due to all three sources is

$$V = 10 + 5 + 2$$
$$= 17 \text{ V}$$

(b) To find the voltage due to one 16 V source alone set the other to zero. The 3 Ω resistor is then in parallel with a 6 Ω resistor giving an equivalent resistance of 2 Ω. The 2 Ω appears in series with a 6 Ω resistor, so by the voltage divider rule one quarter of the source voltage, or 4 V, appears across it. The same calculation applies to either of the voltage sources taken alone so the total voltage V is 8 V.

(c) To find the voltage due to any of the 10 V sources, set the other three to zero. There are then four 8 Ω resistors in parallel giving an effective resistance of 2 Ω.

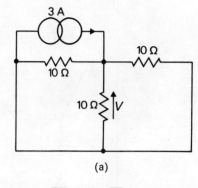

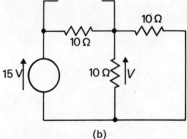

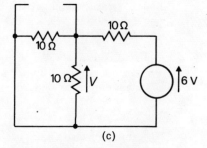

Figure 8 The circuits formed by setting all but one source in figure 7(a) to zero.

The 2 Ω is in series with 8 Ω so the voltage across it is one fifth of the source voltage of 2 V. This calculation applies to all four voltage sources so the total voltage V is $2 \times 4 = 8$ V.

SAQ 1

The linear network in figure 9 has the three input waveforms shown applied to it. If the output waveforms for the first two inputs are as shown, make a rough sketch of the output for the third input.

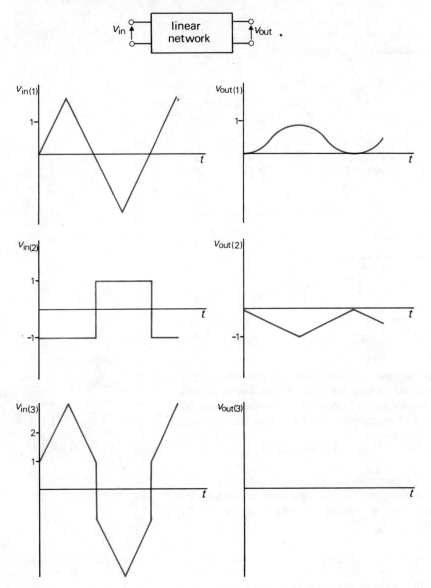

Figure 9

Section 3

7.3 Sinusoidal response of linear components

If we know the response of a linear network to a few signals we can use the principle of homogeneity to find its response to any other signal which is one of the original signals multiplied by a constant (i.e. scaled up or down). The principle of superposition enables us to find the response to any signal formed by adding two or more of the original signals. But what we are really after is the response of a linear network to practically any signal.

Suppose we had a set of 'building block' signals, that is, a set of signals with the property that by adding together appropriate combinations of these signals, we could form any arbitrary signal. Then suppose we found the response of some linear circuit to each of those building blocks. Then knowing only the responses to the building block signals we could find the response to any arbitrary signal by simply adding the responses of the appropriate combination of building blocks.

7.3.1 Sinusoidal signals

There are many possible sets of signals which can act as building blocks in the sense given above. The most widely used are the *sinusoids**.

sinusoids

Any sinusoidal signal can be described by either of two expressions

$$A \cos(\omega t + \phi)$$
$$A \sin(\omega t + \phi)$$

For example, the sinusoid in figure 10 can be written as $V_s \cos t$ or $V_s \sin(t - \pi/2)$.

Let us get our terminology straight as to what a sinusoid is. Its *amplitude* is the height of the peak of the waveform above its average value. Its *peak to peak amplitude* is twice its amplitude. The amount of time after which it repeats (i.e. the time between successive peaks) is its *period*. Its *frequency* is the number of periods (or cycles) in one second. If T is the period (in seconds) and f is the frequency, then

amplitude
peak to peak amplitude

period   frequency

$$f = 1/T$$

The unit of frequency is the hertz (Hz). It has the dimensions of s^{-1}. The mathematical functions $\sin x$ and $\cos x$ repeat after 2π radians. Since the signals $\sin(\omega t + \phi)$ and $\cos(\omega t + \phi)$ repeat after T seconds,

$$\omega T = 2\pi \quad \text{or}$$
$$\omega = \frac{2\pi}{T} = 2\pi f$$

We call ω the *angular frequency* of the sinusoid. Later in this unit I will often use the term 'frequency' to refer to either ω or f. The context will indicate which is meant. The only time it is really essential to distinguish between the two is in numerical examples, where you need to know when to include the factor 2π.

angular frequency

In figure 10 what is the amplitude, peak to peak amplitude, frequency, period and angular frequency of the sinusoid shown?

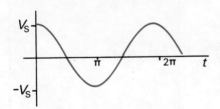

Amplitude: V_s
Peak to peak amplitude: $2V_s$
Period: 2π s
Frequency: $1/2\pi$ Hz
Angular frequency: 1 rad s^{-1}

Figure 10   A sinusoid.

---

* *The correct technical use of the term 'signal' is for varying voltages and currents which convey information. However, in colloquial usage sinusoidal waveforms are often referred to as signals as well.*

I still haven't explained the constant ϕ in the expressions for a sinusoid. It represents the *phase* of the sinusoid.

phase

If two sinusoids have the same frequency but have peaks which occur at different times, as in figure 11, they are said to have a *phase difference*. A phase difference, in radians, is that fraction of the 2π radians in a complete cycle represented by the difference between the occurrences of the two peaks. By describing phase differences as fractions of a cycle, rather than as the actual time between the peaks of the two sinusoids it becomes independent of frequency. There is a phase difference of $\pi/2$ between the two sinusoids in figure 11.

phase difference

The phase of the sinusoid whose peak occurs *earlier* (to the left in figure 11) is said to *lead* that of the other, which is said to *lag*.

leading and lagging phases

Now, if some voltages or currents in a circuit vary sinusoidally how can we determine what formulas to use to describe them? That is, are they $A \cos \omega t$ or $A \sin \omega t$, or if we use the more general formulas, $A \cos (\omega t + \phi)$ or $A \sin (\omega t + \phi)$, what is ϕ?

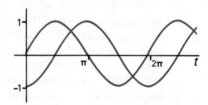

Normally one voltage or current waveform is arbitrarily chosen to be a reference (it is often a source waveform) and it is arbitrarily assigned a formula. The formulas for all other sinusoids in the circuit are then determined by finding their phase difference from the reference sinusoid.

Figure 11 Two sinusoids with a phase difference of $\pi/2$.

In this course unit it will be most convenient to represent the reference sinusoid in all circuits by an expression of the form $A \cos \omega t$. Then whenever I refer to the 'phase' of a sinusoid I mean the phase difference between that sinusoid and a reference described by $A \cos \omega t$.

The phase of a sinusoid which leads $\cos \omega t$ is *positive*, the phase of a sinusoid which lags $\cos \omega t$ is *negative*.

positive and negative phase

What are the phases of $\cos \omega t$, $\sin \omega t$, $\cos (\omega t + \phi)$ and $\sin (\omega t + \phi)$?

Cos ωt is the reference so its phase is zero. Sin ωt lags by $\pi/2$ so its phase is $-\pi/2$. Cos $(\omega t + \phi)$ leads by ϕ so its phase is ϕ. Sin $(\omega t + \phi)$ lags by $\pi/2 - \phi$ (alternatively, it leads by $-\pi/2 + \phi$) so its phase is $-\pi/2 + \phi$.

Since all sinusoids repeat every 2π radians, two sinusoids with a phase difference of 2π, 4π, 6π, etc., can also be described as having zero phase difference. To avoid such ambiguities phase differences are always described as lying between $-\pi$ and π radians.

What is the phase of $-\cos \omega t$?

If you compare the graphs of $\cos \omega t$ and $-\cos \omega t$ in figure 12, you can see that they have a phase difference of $+\pi$ or $-\pi$ radians. Again, to avoid ambiguities, we usually use $+\pi$. So a negative sign in front of a sinusoid is equivalent to a phase difference of π. That is

$$-\cos \omega t = \cos (\omega t + \pi)$$

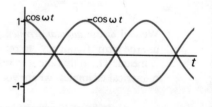

Another useful conversion formula is

$$\sin (\omega t + \phi) = \cos (\omega t + \phi - \pi/2)$$

Figure 12 The graphs of $\cos \omega t$ and $-\cos \omega t$. They have a phase difference of π.

7.3.2 Sinusoids in electronic components and circuits

Now let us see how sinusoidal signals are processed by resistors, capacitors and inductors.

Resistor. If the current is

$$i = I \cos \omega t$$

the voltage is

$$v = Ri$$
$$= RI \cos \omega t$$

Capacitor. If the voltage is

$$v = V \cos \omega t$$

the current is

$$i = C \frac{dv}{dt}$$
$$= -\omega C V \sin \omega t$$
$$= \omega C V \cos (\omega t + \pi/2)$$

Inductor. If the current is

$$i = I \cos \omega t$$

the voltage is

$$v = L \frac{di}{dt}$$
$$= -\omega L I \sin \omega t$$
$$= \omega L I \cos (\omega t + \pi/2)$$

For all three types of components, if either the voltage or current is a sinusoid the other must also be a sinusoid with the same frequency.

Contrast this with the step responses of the *RL* and *RC* circuits in Unit 6. The outputs there were exponentials so the forms of the inputs and outputs were different.

The differences between the responses of the three components are:

1. The current and voltage in a resistor have the same phase. In a capacitor or inductor the current and voltage have a phase difference of $\pi/2$.

2. In a capacitor, the phase of the voltage lags the phase of the current (i.e. the voltage reaches its peak later than the current), while in an inductor it leads the phase of the current.

3. The ratio of the amplitude of the voltage to the amplitude of the current is:

    Resistor: R
    Capacitor: $1/\omega C$
    Inductor: ωL

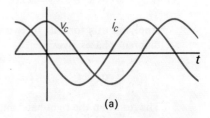

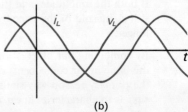

Figure 13 The relative phases for sinusoidal voltages and currents (a) in a capacitor (b) in an inductor.

It is constant for the resistor, but varies with frequency for the capacitor and inductor. Figure 14 shows these three ratios plotted against frequency. The use of logarithmic scales makes all three straight lines. The inductor curve has positive slope. The capacitor curve has the same slope, but negative.

4. The capacitor current is proportional to the rate of change of its voltage. As the frequency increases, the voltage varies more rapidly and so the current increases relative to the voltage. As the frequency gets larger and larger and approaches infinity, the voltage amplitude approaches zero for any current amplitude. At the other extreme, as the frequency gets smaller and smaller and approaches zero, the current amplitude approaches zero for any voltage amplitude.

    The behaviour of a capacitor approaches that of a short circuit as the frequency approaches infinity and approaches that of an open circuit as the frequency approaches zero.

5. The inductor voltage is proportional to the rate of change of the current. Thus as the frequency of the current increases, the voltage increases relative to the current. The situation is just the reverse of that with the capacitor.

    The behaviour of an inductor approaches that of an open circuit as the frequency approaches infinity and approaches that of a short circuit as the frequency approaches zero.

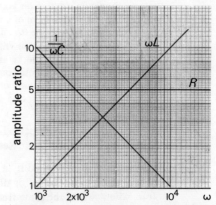

Figure 14 The ratio of the amplitudes of a sinusoidal current and voltage in a resistor, capacitor and inductor as a function of frequency. (Values used for the plot were $R = 5\,\Omega$, $C = 10^{-4}$ F, $L = 10^{-3}$ H.)

If you keep these results at the two extremes of frequency in mind it will help you to have a good understanding of how capacitors and inductors process sinusoidal signals of different frequencies.

If a sinusoidal current of amplitude 1 A flows through a capacitor of 10^{-4} F, and an inductor of 10^{-4} H, what is the amplitude of the voltage across each if the angular frequency is (a) 100 rad s^{-1} or (b) 10^6 rad s^{-1}?

Capacitor voltage amplitude equals 100 V when $\omega = 100$ and 0.01 V when $\omega = 10^6$.
Inductor voltage amplitude equals 0.01 V when $\omega = 100$ and 100 V when $\omega = 10^6$.

Now what will happen if we take our three basic components and connect them together in various combinations? By Kirchhoff's laws the current and voltage in each component will be found by adding or subtracting the currents and voltages in components connected to it. In the TV programme associated with this unit you can see demonstrated that when two sinusoidal voltages with the same frequency are added electronically their sum is also a sinusoidal voltage with the same frequency, no matter what the relative phases or amplitudes of the two added sinusoids.

We can show this property mathematically by using the trigonometric identity

$$\cos x + \cos y = 2 \cos \frac{x-y}{2} \cos \frac{x+y}{2}$$

Applying this formula to two sinusoidal signals with the same amplitude but different phases gives

$$\cos(\omega t + \phi_1) + \cos(\omega t + \phi_2) = 2 \cos \frac{(\phi_1 - \phi_2)}{2} \cos \left(\omega t + \frac{\phi_1 + \phi_2}{2}\right)$$

The result on the right-hand side is a sinusoid. The factor $2 \cos(\phi_1 - \phi_2)/2$ does not vary with time and is the amplitude. From the factor $\cos\{\omega t + (\phi_1 + \phi_2)/2\}$ we can see that the phase is $(\phi_1 + \phi_2)/2$ and the frequency is still ω.

If both the amplitudes and the phases of two sinusoids are different their sum is still a sinusoid of the same frequency, although the mathematical proof is a bit longer.

By combining our observations as to what happens to sinusoidal signals within the basic linear components and when they are connected together, you can see the primary reason why sinusoidal signals are so widely used as building block signals to characterize linear networks.

If all voltage and current sources in a linear circuit produce sinusoidal signals at the same frequency then all currents and voltages in that circuit will also be sinusoids of that frequency.

To appreciate the implications of this statement either try the following experiment or just picture it in your mind.

Display a sinusoid on your home oscilloscope. Suppose that this sinusoid is the source waveform of a linear network. Change the display by (a) changing the y amplifier scale (to change the apparent amplitude) and (b) changing the x position control (so that if the sinusoid were also visible in its original position there would be an apparent phase difference between the two positions). By appropriate combinations of these two changes you can make the display appear like any other waveform in the circuit. That is what I mean when I say the *form* of a sinusoidal signal is not changed by a linear circuit.

The sinusoid is the only periodic* mathematical function whose form is not changed in all the following cases:

(a) when it is multiplied by a constant;
(b) when it is integrated or differentiated;
(c) when two of the same frequency are added or subtracted.

For any other signal applied to a linear network you must find the form of every current and voltage waveform you need. (Compare, for example, the form of the voltage waveforms in the resistor and the capacitor of the RC circuit when a square wave was applied.) In general this requires the solution of a differential equation, but for a sinusoid you know the form before you start. All you need to do is to find the phase and amplitude. To be sure, that is not a trivial problem. The next few sections tell one way of doing it.

* It is only for functions that repeat themselves periodically that it makes sense to talk about 'two of the same frequency'.

Section 4

7.4 Using complex numbers to handle sinusoids

7.4.1 Phasor notation

You now see that the main problem in finding the response of a linear network to a sinusoidal signal is to find the amplitude and phase of the currents and voltages in which you are interested. To do this a mathematical shorthand known as *phasor notation* is used. You will need to be fairly familiar with complex numbers of the form $x + jy$ (where $j = \sqrt{-1}$)* for the rest of this course unit. If you are unsure of how to handle complex numbers, read Part 2 of *Mathematics for Electronics*.

phasor notation

Let us examine the line OP in figure 15. Its length is A. The angle OP makes with respect to the x axis is θ. Its projection on the x axis, OX, has length $A \cos \theta$. Its projection on the y axis, OY, has length $A \sin \theta$. If we now rotate OP anticlockwise about O, the point P will describe a circle of radius A, and the lengths of OX and OY will vary sinusoidally. This is shown in an animated sequence on the TV programme for this unit.

Suppose OP rotates so that the angle θ increases at a steady rate of ω rad s^{-1}. Then OP will be in the same position every 2π radians and the frequency of rotation, f, will be $\omega/2\pi$ Hz. If OP makes an angle of ϕ rad with the x axis initially, then θ can be expressed as a function of time by the equation

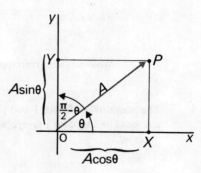

Figure 15

$$\theta = \omega t + \phi$$
$$= 2\pi f t + \phi$$

Therefore the length of each projection is

$$OX = A \cos(\omega t + \phi)$$
$$OY = A \sin(\omega t + \phi)$$

Each projection attains its maximum length at the instant OP passes through the corresponding axis, so the amplitude of the sinusoids is A. Since the rotation is anticlockwise the projection on the x axis reaches its maximum ϕ radians *before* zero, which is consistent with our earlier convention that the phase of $\cos(\omega t + \phi)$ is ϕ. The projection on the y axis reaches its maximum $\pi/2 - \phi$ radians *after* zero so its phase is $-(\pi/2 - \phi)$ or $\phi - \pi/2$ which is also consistent with the earlier convention.

We have now established a correspondence between a line OP and the sinusoids generated by its projections on the x and y axes. To avoid confusion, let us henceforth use only the projection on the x axis. (Sines can always be converted to cosines as shown in the last section.)

To enable us to do such calculations as the sum or difference of two sinusoids we consider P to be a point in the complex plane, represented by the complex number $x + jy$, where

$$x = OX$$
$$y = OY$$

Then instead of adding two sinusoids we can add the corresponding complex numbers. The sum of the two complex numbers then corresponds to the sum of the two sinusoids.

The complex number corresponding to a sinusoid is called the *phasor* of that sinusoid.

phasor

---
* Some mathematicians represent $\sqrt{-1}$ by i for 'imaginary' but since i is used for current in electronics, j is used for $\sqrt{-1}$.

As described in *Mathematics for Electronics* if $x + jy$ is a complex number, x is its *real part* and y is its *imaginary part*. There are two other ways of writing a complex number

$$x + jy = r \angle \phi$$
$$= r \exp j\phi$$

where r is the *magnitude* and ϕ is the *angle* of the complex number.

The magnitude of a phasor equals the amplitude of the corresponding sinusoid. The angle of a phasor equals the phase angle of the corresponding sinusoid.

A phasor can be changed from its real and imaginary form to its magnitude and angle or exponential forms by the formulas

$$r = \sqrt{(x^2 + y^2)}$$

$$\phi = \tan^{-1} \frac{y}{x}$$

$$x = r \cos \phi$$

$$y = r \sin \phi$$

What sinusoids correspond to the phasors of 1 and j?

The magnitude of both 1 and j is 1. The angle of 1 is zero so it corresponds to $\cos \omega t$. The angle of j is $\pi/2$ so it corresponds to $\cos(\omega t + \pi/2)$ which equals $-\sin \omega t$.

What is the phasor for $V \cos(\omega t + \phi)$?

In magnitude and angle form it is $V \angle \phi$. In real and imaginary form it is

$$x + jy = V \cos \phi + jV \sin \phi$$

Now if we think of $x + jy$ as the sum of the two phasors x and jy, we can use the results we found for 1 and j to represent $V \cos(\omega t + \phi)$ as

$$V \cos(\omega t + \phi) = (V \cos \phi) \cos \omega t + (V \sin \phi) \cos\left(\omega t + \frac{\pi}{2}\right)$$

This expression tells us that a sinusoid with an arbitrary phase ϕ equals the sum of a sinusoid with zero phase (the $\cos \omega t$ term) and a sinusoid with a phase of $\pi/2$ (the $\cos(\omega t + \pi/2)$ term). The sinusoid with zero phase is called the *in phase* component and comes from the real part of the phasor. The sinusoid with a phase of $\pi/2$ is called the *out of phase* component and comes from the imaginary part of the phasor.

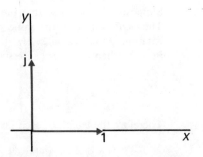

Figure 16 The phasors 1 and j.

in phase and out of phase components

An aside on phasor notation

Some textbooks use a distinct notation, such as **V** for a voltage phasor, to denote phasors or other complex quantities. Such a notation has not been used in this text partly because it is difficult to write without ambiguity and partly because in many circumstances it is not necessary to distinguish these quantities clearly on every occasion. What I have tried to do, and what later units will try to do too, is to point out clearly when phasors are being considered, if this is not obvious from the context.

7.4.2 Exercises on phasor notation*

1. Find the phasors for:

    (a) $10 \cos(\omega t + \pi/4)$
    (b) $5 \sin(\omega t + \pi/3)$
    (c) $12 \cos(\omega t - \pi/3)$
    (d) $\cos(\omega t + \pi)$

    in magnitude and angle form and in real and imaginary form.

2. Find the sinusoids whose phasors are:

    (a) $j2$
    (b) $3 - 3j$
    (c) $1 + j$
    (d) $10 \angle \pi/2$

1. The answers are shown graphically in figure 17.

    (a) In magnitude and angle form the phasor is $10 \angle \pi/4$. In real and imaginary form it is

    $$10 \cos \frac{\pi}{4} + j10 \sin \frac{\pi}{4} = \frac{10}{\sqrt{2}} + j\frac{10}{\sqrt{2}}$$

    (b) $5 \sin\left(\omega t + \frac{\pi}{3}\right) = 5 \cos\left(\omega t + \frac{\pi}{3} - \frac{\pi}{2}\right)$

    $$= 5 \cos\left(\omega t - \frac{\pi}{6}\right)$$

    In magnitude and angle form the phasor is $5 \angle -\pi/6$. In real and imaginary form it is

    $$5 \cos \frac{-\pi}{6} + j5 \sin\left(\frac{-\pi}{6}\right) = \frac{5}{2}\sqrt{3} - j\frac{5}{2}$$

    (c) In magnitude and angle form the phasor is $12 \angle -\pi/3$. In real and imaginary form it is $6 - j6\sqrt{3}$.

    (d) In magnitude and angle form the phasor is $1 \angle \pi$. In real and imaginary form it is -1.

2. (a) Since j corresponds to $-\sin \omega t$, j2 corresponds to $-2 \sin \omega t$.

    (b) In magnitude and angle form the phasor is $3\sqrt{2} \angle -\pi/4$. The sinusoid is $3\sqrt{2} \cos(\omega t - \pi/4)$.

    (c) In magnitude and angle form the phasor is $\sqrt{2} \angle \pi/4$. The sinusoid is $\sqrt{2} \cos(\omega t + \pi/4)$.

    (d) $10 \cos(\omega t + \pi/2) = -10 \sin \omega t$.

7.4.3 Complex impedance

By Ohm's law the ratio of the voltage across a resistor to the current through it at any instant of time is constant, no matter how the current and voltage may vary with time.

The current in a capacitor depends upon the rate of change of the voltage across it. The voltage across an inductor depends upon the rate of change of its current. Thus no constant can characterize the *instantaneous* ratio of voltage to current for inductors or capacitors. However, for a sinusoidal voltage and current in an inductor, for example, the ratio of the phasors of those sinusoids is the same no matter what the current and voltage phasors are. Similarly for capacitors.

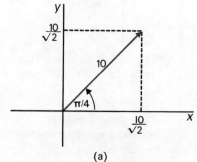
(a)

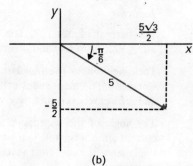
(b)

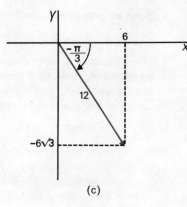
(c)

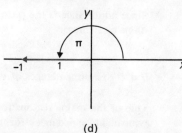
(d)

Figure 17 Answer to section 7.4.2 exercise 1.

---

* *To save time spent on arithmetic remember that $\pi/2$ radians $= 90°$, $\pi/3$ radians $= 60°$, etc. The cosines and sines of these common angles are listed in Mathematics for Electronics.*

The ratio of the voltage phasor to the current phasor is called the *complex impedance* or simply the *impedance* of a component. Using impedances and phasors we can find the response of a circuit containing resistors, capacitors and inductors in a very similar manner to the way we found the dc response of networks containing only resistors.

complex impedance

> Find the impedance of a resistor, inductor and capacitor from the expressions for their sinusoidal response in section 7.3. First find the phasors for the voltage and current, then divide them. Remember that it is easiest to divide complex numbers when they are in magnitude and angle form.

Resistor

The current waveform $I \cos \omega t$ has a phasor I. The voltage waveform $RI \cos \omega t$ has a phasor RI. Their ratio R is the impedance of the resistor.

Capacitor

The voltage phasor is V. The current is $\omega C V \cos(\omega t + \pi/2)$. The current phasor is $\omega C V \angle \pi/2$. The impedance is $1/\omega C \angle -\pi/2$. In real and imaginary form this is $-j/\omega C$ or $1/j\omega C$.

Inductor

The current phasor is I. The voltage phasor is $\omega L I \angle \pi/2$. The impedance is $\omega L \angle \pi/2$, or in real and imaginary form, $j\omega L$.

The magnitude of the impedance is the amplitude ratio of the current and voltage sinusoids. The magnitudes of the capacitor and inductor impedances are $1/\omega C$ and ωL respectively and they vary with frequency as shown in figure 14.

The angle of the impedance is the phase difference between the voltage and current sinusoids. It is $-\pi/2$ and $\pi/2$ for the capacitor and the inductor respectively since the capacitor voltage lags its current and the inductor voltage leads its current.

Since an impedance is the ratio of a voltage to a current it is measured in ohms.

Complex admittance

Occasionally it is useful to use the conductance G of a resistor, rather than its reciprocal R. Similarly it is sometimes more convenient to use the reciprocal of impedance which is called *complex admittance* or just *admittance*. The admittances of the three basic elements are

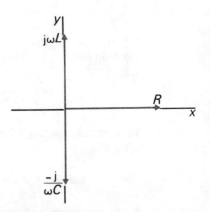

Figure 18 *The impedances of a resistor, capacitor and inductor.*

complex admittance

Resistor: $G (= 1/R)$

Capacitor: $j\omega C$

Inductor: $1/j\omega L \left(= \dfrac{-j}{\omega L} \right)$

Since admittance is the ratio of current to voltage it is measured in reciprocal ohms or siemens.

7.4.4 The impedances of combinations of components

One of the major reasons for using impedances is that with them we can find equivalent impedances for mixed combinations of resistors, capacitors and inductors connected in series and parallel by the use of the same rules as resistors alone were combined in units 1 and 6.

Two components whose complex impedances are Z_1 and Z_2 connected in series are equivalent to an impedance Z_s equal to their sum

impedances of series and parallel connections

$$Z_s = Z_1 + Z_2$$

The same components connected in parallel are equivalent to an impedance Z_p equal to their product divided by their sum

$$Z_p = \frac{Z_1 Z_2}{Z_1 + Z_2}$$

Use the principle of duality to find the formulas for the series and parallel combinations of two elements represented by their admittances Y_1 and Y_2.

For a parallel connection, the formula for the equivalent admittance Y_p is the same as the formula for an equivalent impedance of elements connected in series.

$$Y_p = Y_1 + Y_2$$

For a series connection, the equivalent admittance formula is the same as the equivalent impedance formula for a parallel connection

$$Y_s = \frac{Y_1 Y_2}{Y_1 + Y_2}$$

We can do far more with impedances and admittances than find equivalents for series and parallel combinations of components.

If (a) all currents and voltages in a linear circuit are sinusoids of the same frequency and are represented by their phasors and (b) all components are represented by their impedances (or admittances), then Kirchhoff's laws and the voltage and current divider rules can be used to find currents and voltages in just the same way as for resistive networks. You will see how this is done in an example.

7.4.5 A worked example: an *RL* circuit

Let us find the current i and the voltage v_o, in the circuit of figure 19(a). In figure 19(b) I have used capital letters to indicate the phasors of the current and voltages (because phasors do not change with time) and have indicated the impedances of the components. The phasor V of the source voltage $V_s \cos \omega t$ is

$$V = V_s + j0$$
$$= V_s$$

The equivalent series impedance of the resistor and inductor is found by adding the impedance of the resistor and the inductor. Thus

$$Z_s = R + j\omega L$$

We can use Z_s to find the current phasor because $Z_s = V/I$. Thus

$$I = \frac{V}{Z_s} \qquad (1)$$
$$= \frac{V}{R + j\omega L}$$

The expression for the current phasor $V/(R + j\omega L)$ can be written as a complex number in real and imaginary form as

$$\frac{V}{R + j\omega L} = \frac{V_s + j0}{R + j\omega L} \times \frac{R - j\omega L}{R - j\omega L}$$
$$= \frac{RV_s}{R^2 + \omega^2 L^2} - j\frac{\omega L V_s}{R^2 + \omega^2 L^2}$$

To find the sinusoid which the phasor I represents we need the magnitude and angle of I. The expressions for the magnitude and angle are simpler if found directly from equation (1) rather than from the real and imaginary form of I.

The magnitude of I is written as $|I|$. It is found by dividing the magnitude of the numerator by the magnitude of the denominator of $V/(R + j\omega L)$.

$$|V| = V_s$$
$$|R + j\omega L| = \{R^2 + (\omega L)^2\}^{1/2}$$

Thus

$$|I| = \frac{V_s}{(R^2 + \omega^2 L^2)^{1/2}}$$

The angle of I, which we will call ϕ, is found by subtracting the angle of the denominator from the angle of the numerator.

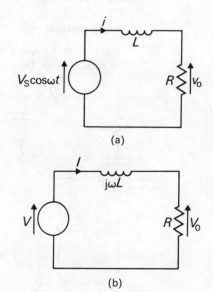

Figure 19 (a) *An RL circuit with a sinusoidal source;* (b) *the same circuit with the phasors of all currents and voltages, and the impedances of the components indicated.*

$$\phi = \text{angle of } V - \text{angle of } (R+j\omega L)$$

$$= 0 - \tan^{-1}\frac{\omega L}{R}$$

Thus the actual current i is given by

$$i = |I|\cos(\omega t + \phi)$$

The phasor of the output voltage V_0 can be found from the voltage divider rule. It is that fraction of the source voltage phasor V given by the ratio of the resistor impedance R and the sum of the inductor impedances $R+j\omega L$.

$$V_0 = \frac{R}{R+j\omega L} V \qquad (2)$$

Again, this expression can be put into real and imaginary form. However, to find the sinusoid v_0 we need the magnitude and angle of V_0, which can most easily be found directly from equation (2).

$$|V_0| = \frac{|R||V|}{|R+j\omega L|} = \frac{RV_s}{(R^2+\omega^2 L^2)^{1/2}}$$

$$\phi_0 = \text{angle of } RV - \text{angle of } (R+j\omega L)$$

$$= 0 - \tan^{-1}\frac{\omega L}{R}$$

Therefore

$$v_0 = |V_0|\cos(\omega t + \phi_0)$$

7.4.6 Exercises on impedances and phasors

Study comment

On your first reading of the unit try only the first two or three parts of each exercise. Do the rest later if you have time.

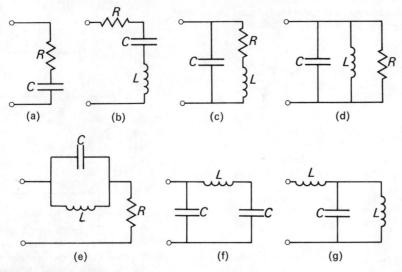

Figure 20

1. Find an equivalent impedance for each of the combinations of components in figure 20.

2. Find the phasors for the voltage v_0 and the current i_0 in each circuit in figure 21.

3. Find the amplitude and phase of the voltage v_0 and current i_0 in each circuit in figure 21 (i.e. find the magnitude and angle for each phasor found in 2 above).

Your results should be the same as or equivalent to the following:

1. (a) $R + \dfrac{1}{j\omega C}$

   (b) $R + j\omega L + \dfrac{1}{j\omega C} = R + j\left(\omega L - \dfrac{1}{\omega C}\right)$

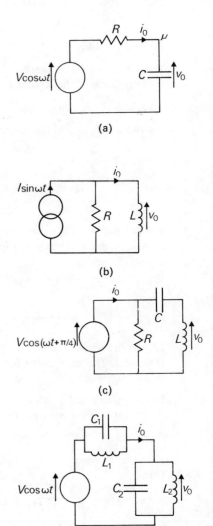

Figure 21

(c) The series R and L give $R+j\omega L$. This combination in parallel with C gives

$$\frac{(1/j\omega C)(R+j\omega L)}{1/j\omega C + R + j\omega L} = \frac{R+j\omega L}{1-\omega^2 LC + j\omega CR}$$

(d) First add the *admittances* of the three components

$$j\omega C + \frac{1}{j\omega L} + \frac{1}{R}$$

The impedance is the reciprocal of this expression

$$\frac{1}{j\omega C + 1/j\omega L + 1/R} = \frac{j\omega LR}{R - \omega^2 LRC + j\omega L}$$

(e) The parallel L and C combination gives

$$\frac{j\omega L(1/j\omega C)}{j\omega L + 1/j\omega C}$$

Including the series R gives

$$R + \frac{j\omega L(1/j\omega C)}{j\omega L + 1/j\omega C} = \frac{R(1-\omega^2 LC) + j\omega L}{1 - \omega^2 LC}$$

(f) The series L and C gives

$$j\omega L + \frac{1}{j\omega C}$$

Including the parallel C gives

$$\frac{(j\omega L + 1/j\omega C)1/j\omega C}{j\omega L + 1/j\omega C + 1/j\omega C} = \frac{j(\omega L - 1/\omega C)}{2 - \omega^2 LC}$$

(g) The series L is added to the parallel C and L to give

$$j\omega L + \frac{j\omega L(1/j\omega C)}{j\omega L + 1/j\omega C} = \frac{j\omega L(2-\omega^2 LC)}{1-\omega^2 LC}$$

2 (a) The source phasor is $V + j0$. Then

$$I_0 = \frac{V}{R + 1/j\omega C} = \frac{j\omega CV}{1 + j\omega CR}$$

By the voltage divider rule

$$V_0 = \frac{(1/j\omega C)}{1/j\omega C + R} V$$

$$= \frac{V}{1 + j\omega CR}$$

(b) The source phasor is $-jI$. By the current divider rule

$$I_0 = \frac{R(-jI)}{R + j\omega L}$$

$$= \frac{-jRI}{R + j\omega L}$$

$$V_0 = (j\omega L)I_0$$

$$= \frac{\omega LRI}{R + j\omega L}$$

(c) The source phasor is $V \angle \pi/4$ or $V/\sqrt{2} + jV/\sqrt{2}$. By the voltage divider rule

$$V_0 = \frac{j\omega L}{j\omega L + 1/j\omega C}\left(\frac{V}{\sqrt{2}} + \frac{jV}{\sqrt{2}}\right)$$

$$= \frac{-\omega^2 LC}{1 - \omega^2 LC}\left(\frac{V}{\sqrt{2}} + \frac{jV}{\sqrt{2}}\right)$$

To find I_0 first find the total admittance. The admittance of the series L and C is

$$\frac{j\omega C(1/j\omega L)}{j\omega C + 1/j\omega L} = \frac{j\omega C}{1 - \omega^2 LC}$$

Adding the parallel R gives

$$\frac{1}{R} + \frac{j\omega C}{1 - \omega^2 LC}$$

The current phasor now is

$$I_0 = \left(\frac{1}{R} + \frac{j\omega C}{1 - \omega^2 LC}\right)\left(\frac{V}{\sqrt{2}} + \frac{jV}{\sqrt{2}}\right)$$

(d) The source phasor is $V + j0$. The total impedance is the sum of the impedances of each parallel L and C, or

$$\frac{j\omega L_1(1/j\omega C_1)}{j\omega L_1 + 1/j\omega C_1} + \frac{j\omega L_2(1/j\omega C_2)}{j\omega L_2 + 1/j\omega C_2}$$

$$= \frac{j\omega L_1}{1 - \omega^2 L_1 C_1} + \frac{j\omega L_2}{1 - \omega^2 L_2 C_2}$$

Thus

$$I_0 = \frac{V}{j\{\omega L_1/(1 - \omega^2 L_1 C_1) + \omega L_2/(1 - \omega^2 L_2 C_2)\}}$$

By the current divider rule, the current through L_2 is

$$I_{L2} = \frac{1/j\omega C_2}{1/j\omega C_2 + j\omega L_2} I_0$$

$$= \frac{I_0}{1 - \omega^2 L_2 C_2}$$

Then

$$V_0 = j\omega_2 I_{L2}$$

$$= \frac{j\omega L_2 I_0}{1 - \omega^2 L_2 C_2}$$

3 To find magnitudes we divide the magnitude of the numerator by the magnitude of the denominator. We find angles by subtracting the angle of the denominator from the angle of the numerator. Let us denote the angle of V_0 by ϕ_V and the angle of I_0 by ϕ_I.

(a) $|I_0| = \dfrac{\omega CV}{\{1 + (\omega RC)^2\}^{1/2}}$

$\phi_I = \dfrac{\pi}{2} - \tan^{-1} \omega RC$

$|V_0| = \dfrac{V}{\{1 + (\omega RC)^2\}^{1/2}}$

$\phi_V = -\tan^{-1} \omega RC$

(b) $|I_0| = \dfrac{RI}{\{R^2 + (\omega L)^2\}^{1/2}}$

$\phi_I = -\dfrac{\pi}{2} - \tan^{-1} \dfrac{\omega L}{R}$

$|V_0| = \dfrac{\omega LRI}{\{R^2 + (\omega L)^2\}^{1/2}}$

$\phi_V = -\tan^{-1} \dfrac{\omega L}{R}$

(c) V_0 is simply a real expression times the source phasor. Thus

$$V_0 = \left|\frac{\omega^2 LC}{1 - \omega^2 LC}\right| V$$

ϕ_V (the phase of the voltage phasor) is either $\pi/4$ or $\pi/4 - \pi$, the latter if $-\omega^2 LC/(1 - \omega^2 LC)$ is negative.

Since the current phasor is the product of two terms we multiply the magnitudes and add the angles.

$$|I_0| = V\left\{\frac{1}{R^2} + \left(\frac{\omega C}{1-\omega^2 LC}\right)^2\right\}^{1/2}$$

$$\phi_I = \tan^{-1}\frac{\omega CR}{1-\omega^2 LC} + \frac{\pi}{4}$$

(d) $\quad |I_0| = \dfrac{V}{\omega L_1/(1-\omega^2 L_1 C_1) + \omega L_2/(1-\omega^2 L_2 C_2)}$

$$\phi_I = -\frac{\pi}{2}$$

For V_0, again we have a product

$$|V_0| = \left|\frac{j\omega L_2}{1-\omega^2 L_2 C_2}\right| \times |I_0|$$

$$= \frac{\omega L_2}{1-\omega^2 L_2 C_2}\left(\frac{V}{\omega L_1(1-\omega^2 L_1 C_1) + \omega L_2(1-\omega^2 L_2 C_2)}\right)$$

$$\phi_V = \text{angle of } \frac{j\omega L_2}{1-\omega^2 L_2 C_2} + \phi_I$$

$$= \frac{\pi}{2} - \frac{\pi}{2}$$

$$= 0$$

7.4.7 Summary of sections 7.3 and 7.4

These two sections are the most important in the course unit. Section 7.3 describes what sinusoids are, and the relationships between sinusoidal voltages and currents in linear components. Section 7.4 describes the mathematical techniques, phasors and complex impedance, used to represent sinusoids and their relations in linear components.

A voltage or current waveform which can be represented by $A\sin(\omega t + \phi)$ or $A\cos(\omega t + \phi)$ is a sinusoid. Its amplitude is A, its angular frequency is ω, its frequency is $\omega/2\pi$. Its phase difference with respect to a reference sinusoid represented by $\cos\omega t$, is ϕ for $A\cos(\omega t + \phi)$ and $\phi - \pi/2$ for $A\sin(\omega t + \phi)$.

A phasor is a complex number used to represent the phase and amplitude of a sinusoid. The magnitude of the phasor equals the amplitude of the sinusoid. The angle of the phasor equals the phase of the sinusoid. In real and imaginary form, the real part of the phasor equals the amplitude of the in phase component. The imaginary part of the phasor equals the amplitude of the out of phase component.

The formulas for converting from $x + jy$ form to $r \angle \phi$ form are

$$x = r\cos\phi \qquad y = r\sin\phi$$

$$r = (x^2 + y^2)^{1/2} \qquad \phi = \tan^{-1}\frac{y}{x}$$

The ratio of the amplitude of the voltage to the amplitude of the current in a linear component is

Resistor: $\quad R$

Capacitor: $\quad \dfrac{1}{\omega C}$

Inductor: $\quad \omega L$

The voltage and current in a resistor are in phase. The voltage in a capacitor lags the current by $\pi/2$. The voltage in an inductor leads the current by $\pi/2$.

If the waveforms of all sources in a linear circuit are sinusoids of the same frequency then all other currents and voltages in that circuit are sinusoids of that frequency.

The impedance of a component or a combination of components is the ratio of the phasor of its voltage to the phasor of its current.

The impedances of linear components are

    Resistor: R

    Capacitor: $\dfrac{1}{j\omega C}$

    Inductor: $j\omega L$

The impedances of resistors, capacitors and inductors can be combined using the same rules as for combining resistors.

With all voltages and currents in a circuit represented by their phasors and all components represented by their impedances, Kirchhoff's laws and the voltage and current divider rules can be used to find the voltage and current phasors.

Section 5

7.5 Frequency characteristics of circuits

In section 7.4 we saw how to use phasors to represent sinusoids and how to use complex impedance to find the phasors of any voltage or current in a circuit. The expressions we derived in the example and exercises described the sinusoidal response of various circuits. In the last section we concentrated on the techniques used to find these expressions. Now we will examine the properties of these expressions at different frequencies.

Remember that sinusoids are building blocks which enable us to find the response of a linear circuit to any signal. Different signals can be formed by adding appropriate sinusoids of different frequencies. (The sum of two sinusoids is a sinusoid *only* if their frequencies are the same.) Those sinusoids which can be added to form a given signal are called the *frequency components* of that signal. Thus it is necessary to know the response of a network to sinusoids of all frequencies likely to be found as components of any signal we want to apply to that network.

frequency components

For example, a medium wave radio receiver must accept signals with frequency components in the range 500–1600 kHz. However, at any one setting of the tuning knob it must accept only a very limited band of frequencies within that range and reject all others. In that way it can select a given station. Circuits such as the one within the radio receiver which allow only certain ranges of frequency components to pass are called frequency selective circuits or *filters*. A range of frequencies which is permitted to pass through is called a *pass band*. A range which is stopped is called a *stop band*.

filters
pass band
stop band

In this section we will see how some of the circuits we have already met, *RC*, *RL* and *RLC* circuits, can be used as frequency selective circuits.

7.5.1 Describing frequency characteristics

Before looking at the details of the frequency response of a circuit let us determine the kind of information we want and how best to describe it. For the *RL* circuit in figure 19, which we used as an example, we found an expression for the phasor of the current by dividing the phasor for the source voltage by the impedance.

$$I = \frac{V}{R + j\omega L}$$

The phasor of the source represents only the amplitude and phase of the source voltage and not the frequency. The phasor of the current also depends upon frequency because the impedance depends upon frequency. The information describing how the circuit affects the current is contained in the impedance, the *ratio* of the voltage to the current phasors, rather than in the expression for the current phasor itself. (Or, equally well, it is contained in the admittance of the circuit.)

The expression we found for the phasor of V_0 was

$$V_0 = \frac{R}{R + j\omega L} V$$

Again, the information describing the effect of the circuit is contained in the voltage ratio

$$\frac{V_0}{V} = \frac{R}{R + j\omega L}$$

rather than in the expression for the voltage phasor itself.

In general, the expressions used to characterize the sinusoidal response of circuits are ratios of phasors rather than expressions for phasors themselves.

These ratios may either be impedances or admittances, or voltage or current ratios. The ratio of a voltage measured in one part of a circuit to a source current measured at some other part is called a *transfer impedance*. The ratio of a current measured in one part of a circuit to a source voltage measured at some other part is called a *transfer admittance*. The general name for any of these ratios of phasors is *transfer function*.

transfer impedance

transfer admittance
transfer function

A transfer function is a complex* expression. We can represent it graphically by plotting two graphs, one showing its magnitude against frequency and the other showing its angle against frequency. These two graphs are called the *frequency response curves* of a circuit.

frequency response curves

An alternative graph called a *polar locus plot*, is sometimes used to depict transfer functions. For every value of ω (or f), the transfer function represents some point in the complex plane. The polar locus plot is a graph of these points for all values of ω.

polar locus plot

Logarithmic scales and decibel notation

The range of values over which frequencies or magnitudes may vary in a frequency response curve is often inconveniently large. For example, if we are concerned with the response of a high fidelity amplifier over a frequency range of 30–15 000 Hz, that part of the curve corresponding to the bass notes in music (30–200 Hz) will be squeezed into 1% of the total graph in a linear plot of magnitude against frequency. To overcome this problem we often plot the variables on logarithmic graph paper.

On a logarithmic scale the space allocated to each decade is the same. Thus the range 10–100 Hz has the same space as that allocated to 100–1000 Hz and 1000–10 000 Hz.

There is a special unit, the *decibel* (dB) used to describe ratios of power. It is also used to plot ratios of currents or voltages in logarithmic form.

decibel

If V_1 and V_2 are the phasors of two voltages, then their ratio, in decibels, is defined as

$$\text{voltage ratio in dB} = 20 \lg \frac{|V_1|}{|V_2|}$$

(lg represents $\log_{10}$, whereas $\log_e$ is represented by ln.)

> The term decibel originated in the measurement of sound energy, because the ear responds to changes in sound level logarithmically. The unit used to measure the ratio of two energy levels was the bel (after Alexander Graham Bell). The ratio of two energies E_1 and E_2 in bels is
>
> $$\text{energy ratio in bels} = \lg \frac{E_1}{E_2}$$
>
> This turned out to be too large for most purposes, so *deci*bels were used instead. In electrical terms, the corresponding usage of decibels is to compare powers. If P_1 and P_2 are two values of power, their ratio in dB is
>
> $$\text{power ratio in dB} = 10 \lg \frac{P_1}{P_2}$$
>
> If P_1 and P_2 represent the power in the *same* resistor R and correspond to voltages of magnitude $|V_1|$ and $|V_2|$ then
>
> $$\text{dB} = 10 \lg \frac{|V_1|^2/R}{|V_2|^2/R}$$
>
> $$= 10 \lg \frac{|V_1|^2}{|V_2|^2}$$
>
> $$= 20 \lg \frac{|V_1|}{|V_2|}$$

* *'Complex' in the sense of 'complex' number, not 'complicated'.*

The last expression is the one I used as my definition of a decibel. Although it is equivalent to the historical definition of a decibel *only* when comparing two voltages or currents in the same impedance (or two impedances of the same value) it is widely used for any voltage or current ratio. However, when using decibel notation to compare voltages or currents in different impedances, the impedances should always be mentioned so that incorrect conclusions about the power levels won't be drawn.

What is the ratio in decibels of voltages whose phasors are: (a) V and $10V$; (b) V and $2V$?

(a) $\quad 20 \lg \dfrac{10|V|}{|V|} = 20 \text{ dB}$

(b) $\quad 20 \lg \dfrac{2|V|}{|V|} = 20 \lg 2$

$\qquad \simeq 6 \text{ dB}$

These are useful figures to remember. A voltage increase or decrease by a factor of 10 gives a 20 dB change, while a voltage doubling or halving gives a 6 dB change.

What voltage or current ratio corresponds to a halving or doubling of the power in a resistance?

If a signal V_2 produces twice the power of V_1 in a load R then

$$\dfrac{|V_2|^2/R}{|V_1|^2/R} = 2$$

So

$$|V_1| = \dfrac{|V_2|}{\sqrt{2}}$$

$$= 0.707 |V_2|$$

and in decibels

$$10 \lg \dfrac{|V_2|^2}{|V_1|^2} = 10 \lg 2$$

$$\simeq 3 \text{ dB}$$

Thus a change of $+3$ dB in a voltage ratio across an impedance corresponds to a *doubling* in power. A change of -3 dB indicates a *halving* of power (because $-\lg x = \lg 1/x$).

7.5.2 Frequency response of *RL* and *RC* circuits

Let us now complete the analysis of the *RL* circuit in figure 19 which was already examined in the last section. We want to examine its impedance $R + j\omega L$ for different values of ω. Figure 22(a) is an Argand diagram showing $R + j\omega L$. It equals the sum of two terms representing the impedance of the resistor R and of the inductor $j\omega L$. For different values of ω, R remains constant, but $j\omega L$ changes. The impedance is shown for three values of ω in figure 22(b). I have labelled these three values ω_0, ω_1 and ω_2 and have labelled the corresponding impedances Z_0, Z_1 and Z_2. Let us first consider the middle value ω_0. I chose it so that the magnitudes of the two components would be equal

$$|j\omega_0 L| = |R|$$

or

$$\omega_0 = \dfrac{R}{L}$$

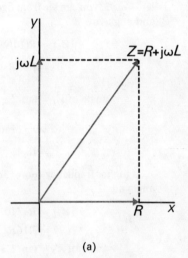

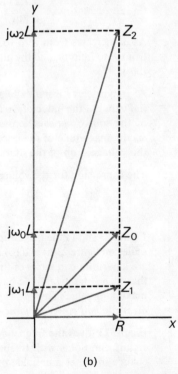

Figure 22 (a) The impedance of the RL circuit; (b) the impedance for three different frequencies.

From the diagram you can see that the angle of Z_0 is 45° or $\pi/4$ rad. At ω_0 neither the resistive nor inductive component predominates. The magnitude of Z_0 is

$$|Z_0| = \{R^2 + (\omega_0 L)^2\}^{1/2}$$
$$= 1.41R$$

The formula for the angle of Z_0 gives

$$\text{angle of } Z_0 = \tan^{-1} \frac{\omega L}{R}$$
$$= \tan^{-1} 1$$
$$= \frac{\pi}{4} \text{ rad}$$
$$= 45°$$

which agrees with the geometric result. The frequency ω_0 is the pivotal frequency since at higher frequencies the inductive component predominates and at lower frequencies the resistive component predominates. At a lower frequency $\omega_1 = \omega_0/3$, you can see from figure 22(b) that Z_1 is much closer to R. Its magnitude and angle are

$$|Z_1| = R(10/9)^{1/2}$$
$$= 1.05R$$

$$\text{angle of } Z_1 = \tan^{-1} \frac{1}{3}$$
$$= 0.32 \text{ rad}$$
$$= 18.4°$$

At a higher frequency ($\omega_2 = 3\omega_0$) Z_2 is much closer to $j\omega L$. Its magnitude and angle are

$$|Z_2| = R\sqrt{10}$$
$$= 3.16R$$
$$\text{angle of } Z_2 = \tan^{-1} 3$$
$$= 1.25 \text{ rad}$$
$$= 71.6°$$

As ω increases from zero to infinity the magnitude of the impedance increases from R to infinity, and the angle of the impedance increases from zero to $\pi/2$ rad (90°).

At frequencies substantially above ω_0 the magnitude of the impedance is essentially that of the inductor, so by the voltage divider rule, most of the voltage drop of the source appears across the inductor. At frequencies substantially below ω_0 the magnitude of the impedance is essentially that of the resistor so most of the voltage drop of the source appears across the resistor.

The expression for the voltage ratio of the RL circuit is

$$\frac{V_0}{V} = \frac{R}{Z}$$

Figure 23 shows the magnitude and angle of V_0/V plotted against frequency. The scales on figure 23(a) and (b) are linear, but on (c) and (d) are logarithmic. (They are logarithmic in terms of the voltage ratio, although linear in decibels.) Notice that at ω_0 the magnitude has changed by a factor of $1/\sqrt{2}$ or -3 dB from its maximum value. Thus ω_0 is called the 3 dB *frequency*. An alternative name for ω_0 is the *half power frequency*.

3 dB frequency
half power frequency

Figure 23 gives the frequency response curves of the output voltage of the RL circuit. Sinusoids with frequencies well below ω_0 pass through with very little change in either amplitude or phase. Sinusoids with frequencies well above ω_0 pass through with their phases changed and with amplitudes attenuated. Thus the circuit is frequency selective and, in particular, is a *low pass circuit*. There is no one frequency below which sinusoids are passed and above which sinusoids

low pass circuit

are blocked because the transition is gradual. But, as you might guess, the frequency which is used (arbitrarily) to designate the boundary between the pass band and the stop band is the 3 dB frequency ω_0. That gives it a third name, the *cut off frequency*. (It has more interpretations than that. Its reciprocal $1/\omega_0$ is L/R, the time constant of the exponential waveform found in Unit 6 when a step was applied to the circuit.)

cut off frequency

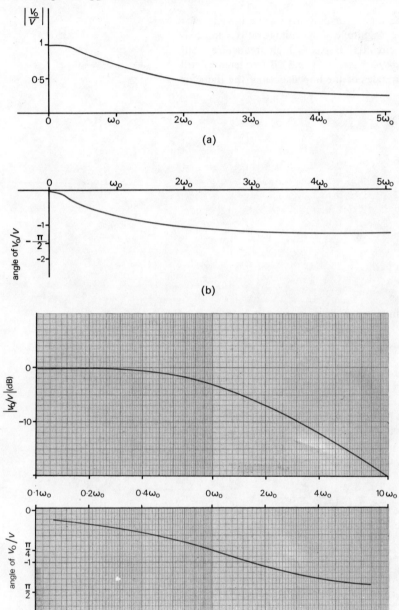

Figure 23 The frequency response of a low pass circuit; (a) and (b) have linear scales; (c) and (d) have logarithmic scales.

From the logarithmic amplitude curve, for frequencies well above ω_0 the amplitude decreases by 6 dB every time the frequency is doubled. This doubling of frequency is called an *octave*, from musical terminology. The amplitude curve is approximately linear in this region with a slope of -6 dB per octave.

What is -6 dB in terms of voltage or current ratios?

It is equivalent to a ratio of $1/2$, so a slope of -6 dB per octave simply means that the amplitude is halved when the frequency is doubled.

Now let us look at a slight variation of the RL circuit, shown in figure 24. The components are the same, but the output voltage is now taken across the inductor.

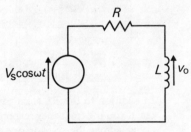

Figure 24

Before reading on, test your understanding of the preceding discussion by answering the following questions about the circuit in figure 24.

1 Will the circuit also be a low pass circuit?
2 What is its 3 dB frequency?

Since the voltage drop of the source appears mainly across the inductor for frequencies substantially above ω_0, the magnitude of the voltage ratio is approximately 1 for high frequencies. The circuit thus passes high frequencies, but attenuates low frequencies. It is a *high pass circuit*. The 3 dB frequency is still ω_0, the frequency at which the magnitudes of the impedances of the inductor and resistor are equal.

high pass circuit

The expression for the voltage ratio is

$$\frac{V_0}{V} = \frac{j\omega L}{Z}$$

Find its magnitude and angle.

$$\left|\frac{V_0}{V}\right| = \frac{\omega L}{\{R^2 + (\omega L)^2\}^{1/2}}$$

$$\text{angle of } \frac{V_0}{V} = \frac{\pi}{2} - \tan^{-1}\frac{\omega L}{R}$$

Let us rewrite these two expressions in terms of ω_0.

$$\left|\frac{V_0}{V}\right| = \frac{\omega/\omega_0}{\{1 + (\omega/\omega_0)^2\}^{1/2}}$$

$$\text{angle of } \frac{V_0}{V} = \frac{\pi}{2} - \tan^{-1}\frac{\omega}{\omega_0}$$

The magnitude and angle of V_0/V are plotted in figure 25. The figure shows that the circuit is indeed a high pass circuit. Notice that for frequencies well below ω_0 the magnitude of V_0/V increases by 6 dB per octave. Well above ω_0 the magnitude is nearly constant at 0 dB, or a voltage ratio of 1.

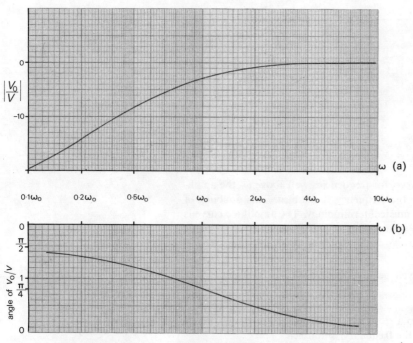

Figure 25 The frequency response of a high pass RL circuit.

In addition to the *RL* circuit we have just discussed, this section is intended to teach you the frequency response of the *RC* circuit with an output voltage taken either across the resistor or the capacitor (figure 26). Since the results are so similar to those of the *RL* circuit I will let you work it out for yourself by setting the circuit as an SAQ.

SAQ 2

Consider the circuits of figure 26.

1. What is the impedance of these circuits? Draw an Argand diagram for the impedance. Find the 3 dB frequency from the Argand diagram as we did for the *RL* circuit. Which element dominates the impedance at frequencies above and below the 3 dB frequency?

2. From this description of the impedance at frequencies above and below the 3 dB frequency determine which of the circuits in figure 26 will be a high pass and which will be a low pass circuit.

3. Find V_o/V, its magnitude and its angle. Sketch the magnitude and angle against frequency. (Hint: substitute the 3 dB frequency into your expressions and compare them with the expressions and curves for the *RL* circuit.)

SAQ 3

The circuit of figure 27 shows an amplifier connected to a 1 kΩ load resistor through a coupling capacitor C. Suppose that the output side of the amplifier can be represented by a sinusoidal voltage source in series with a 1 kΩ resistor.

1. Find an expression in terms of C for the phasor of the voltage v_o across the load resistor.

2. What is the maximum amplitude of the voltage across the load resistor?

3. What is the minimum value of C required to ensure that the ratio of the amplitude of the voltage across the load resistor to its maximum value is never less than -3 dB for all frequencies above 1 kHz?

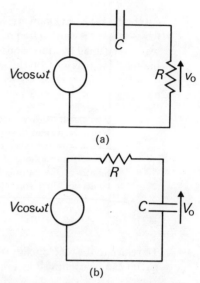

Figure 26 An RC circuit with the output across (a) the resistor; (b) the capacitor.

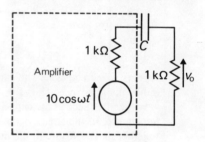

Figure 27

7.5.3 Frequency response of an *RLC* circuit

The circuits we have just discussed each contained just one component with a frequency dependent impedance, either a capacitor or an inductor. We will now discuss the *RLC* circuit in figure 28, which contains both.

Before examining this circuit from the phasor–complex impedance point of view let us try to predict its behaviour in a fairly intuitive way on the basis of our discussion of its step response in Unit 6. We are primarily concerned with the case where the resistor value is substantially below the value required for critical damping so the circuit is underdamped. The response of the circuit to a step in voltage is a sinusoidal oscillation which is exponentially damped. The oscillation is due to the continual interchange of stored magnetic energy in the inductor and stored electric energy in the capacitor. The exponential damping occurs because some energy is removed on each cycle by the resistor.

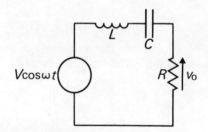

Figure 28

Now suppose that instead of a voltage step, the source supplies a sinusoidal voltage. If the source frequency is the same as the natural frequency of the circuit it will reinforce the natural oscillation so that if R is small a large current will flow. At other frequencies the source will be 'out of step' with the natural frequency of the circuit. Thus it will not reinforce the natural oscillation and so will not produce as large a current. The result is that the output voltage, v_o (which is proportional to the current) will have a maximum amplitude at the natural frequency and will have a smaller amplitude at other frequencies. The circuit will thus behave as a *band pass circuit*.

band pass circuit

This phenomenon of an excitation at just the natural frequency of a system causing a large response is called *resonance*. It is found not only in *RLC* circuits, nor only in electronics for that matter, but in a wide variety of other physical systems. It generally occurs in a system where energy is alternately stored in one of two forms. The pendulum clock is a common example. The energy of the pendulum takes the form of kinetic energy when the pendulum is moving rapidly at the bottom of its swing and takes the form of potential energy at either end of the swing where the pendulum is momentarily still. The escapement mechanism

resonance

ensures that the pendulum is given an extra push to overcome its frictional energy losses at just the natural frequency – one push in each direction per cycle. In this way a very small fraction of added energy is needed to keep the pendulum swinging for a long period of time.

> You can try this yourself with a simple demonstration. Make a pendulum with a weight and some string. Hold the string in one hand and give the weight a push with the other so you can observe the natural frequency. Now stop the weight's motion. Move the hand holding the string back and forth very slightly at the natural frequency. Observe the large resulting swing which builds up. Move your hand at a different rate and notice how much smaller is the swing.

To return to the *RLC* circuit, the prediction that it should behave as a band pass circuit can be confirmed by examining its impedance.

> What is the impedance of the *RLC* circuit at the low and high extremes of frequency?

At very low frequencies the capacitor impedance is very high compared to the inductor impedance. Thus the circuit behaves like an *RC* high pass circuit. At very high frequencies the capacitor impedance is much smaller than the inductor impedance, so the circuit behaves like an *RL* low pass circuit. Both these effects together make the circuit a band pass circuit.

The circuit impedance is

$$Z = R + j\omega L + \frac{1}{j\omega C}$$

$$= R + j\left(\omega L - \frac{1}{\omega C}\right)$$

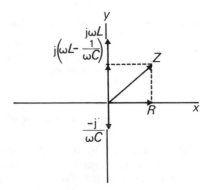

Figure 29

It is shown in the Argand diagram of figure 29. The impedances of the inductor and capacitor are drawn along the y axis. Notice that they are in opposite directions. Their sum, which is

$$j\left(\omega L - \frac{1}{\omega C}\right)$$

is then found and added to the impedance of the resistor R as shown.

Since the magnitude of the inductor impedance increases with increasing frequency and the magnitude of the capacitor impedance decreases with increasing frequency, there is some frequency at which the two magnitudes are the same. The Argand diagram at this frequency is shown in figure 30. Since the capacitor and inductor impedances are in opposite directions, they cancel so the total impedance is simply R. The angle of the impedance is zero. Now since

$$\omega_0 L = \frac{1}{\omega_0 C}$$

we get

$$\omega_0 = \frac{1}{(LC)^{1/2}}$$

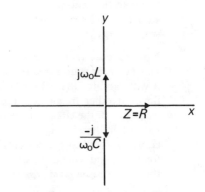

Figure 30 *The Argand diagram for an RLC circuit at resonance.*

Thus the resonant frequency is the same as the natural frequency found in Unit 6.

Since the impedances of the capacitor and inductor cancel at the resonant frequency, the entire source voltage appears across the resistor.

> Does this mean there is no voltage drop across the capacitor or the inductor?

No. By the voltage divider rule the capacitor voltage phasor V_C and the inductor voltage phasor V_L are

$$V_C = \frac{1/j\omega_0 C}{Z} V$$

$$= \frac{1/j\omega_0 C}{R} V$$

$$= \frac{-j}{R}\left(\frac{C}{L}\right)^{1/2} V$$

$$V_L = \frac{j\omega_0 L}{Z} V \qquad (3)$$

$$= \frac{j\omega_0 L}{R} V$$

$$= \frac{j}{R}\left(\frac{L}{C}\right)^{1/2} V$$

Not only are these voltages not zero, but if the factor $(1/R)(L/C)^{1/2}$ is greater than one, they will be larger than the source voltage! However, since they have opposite signs (i.e. are 180° out of phase) the voltage across both in series is zero.

The constant multiplying the voltage is usually denoted by Q for *quality factor*.

quality factor

$$Q = \frac{1}{R}\left(\frac{L}{C}\right)^{1/2}$$

From equation 3, it represents the ratio of the magnitudes of the capacitor impedance (or the inductor impedance) to the total impedance at resonance. It has other interpretations too, as you shall see, which make it one of the basic constants describing a resonant circuit. If Q is large the behaviour of the circuit at frequencies near resonance is primarily determined by the capacitor and inductor, rather than the resistor.

So much for the circuit just at the resonant frequency. Now let us examine its behaviour at higher and lower frequencies. The Argand diagrams in figure 31 show the impedance for two values. Notice that the angle of Z is positive like an RL circuit for frequencies greater than ω_0 and negative like an RC circuit for frequencies below ω_0. The magnitude of Z is greater than at ω_0 in both cases.

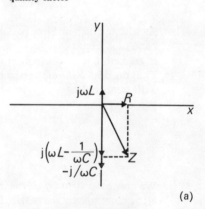

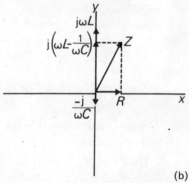

Figure 31 The Argand diagram for an RLC circuit at frequencies (a) below and (b) above the resonant frequency.

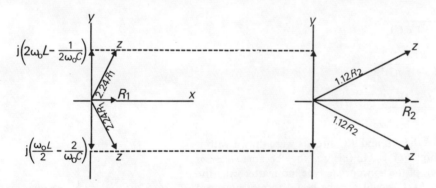

Figure 32 Argand diagrams for two RLC circuits with the same L and C but different R. At frequencies near resonance the change in magnitude of Z is much greater for the smaller value of R.

The resonant frequency is determined by L and C. For given values of L and C the sharpness of the resonance is determined by R. This can be seen by comparing the Argand diagrams in figure 32(a) and (b). Both show the impedance of an RLC circuit at its resonant frequency and twice and one-half the resonant frequency. Notice that the sum of the capacitor and inductor impedances (the reactive components) are the same in (a) and (b). However, the resistor impedance is half the reactive component in (a) and twice the reactive component in (b). Now compare the magnitude change which occurs for each case as the frequency changes from either $2\omega_0$ or $\frac{1}{2}\omega_0$ to resonance, ω_0. In (a), at $2\omega_0$ or $\frac{1}{2}\omega_0$, the magnitude of Z (its length in the diagram) is $2.24 R_1$. At resonance $|Z| = R_1$, a decrease of 55%. In (b), at $2\omega_0$ or $\tfrac{1}{2}\omega_0$, $|Z| = 1.12 R_2$. At resonance $|Z| = R_2$, a decrease of 11%.

The 3 dB frequencies, as with the RL and RC circuits, are those at which the magnitude of the reactive part of the impedance equals the magnitude of the resistive part of the impedance. There are now two such frequencies, one above ω_0 and one below ω_0. For the 3 dB frequency above ω_0

$$\omega_2 L - \frac{1}{\omega_2 C} = R$$

or

$$\omega_2 = \frac{R}{2L} + \left\{ \left(\frac{R}{2L}\right)^2 + \frac{1}{LC} \right\}^{1/2}$$

For the 3 dB frequency below ω_0

$$\frac{1}{\omega_1 C} - \omega_1 L = R$$

or

$$\omega_1 = -\frac{R}{2L} + \left\{ \left(\frac{R}{2L}\right)^2 + \frac{1}{LC} \right\}^{1/2}$$

As with the high and low pass circuits, there is no critical frequency at which the band pass circuit changes from 'stopping' to 'passing' sinusoids. The transition is gradual, but by convention the 3 dB points mark the edges of the pass band. Thus the *band width* of the pass band is the difference between the upper and lower 3 dB frequencies.*

band width

$$W = \omega_2 - \omega_1$$

$$= \frac{R}{L}$$

We now have virtually all the information we need about the RLC circuit. It can all be conveniently summarized in the magnitude and angle curves of the voltage ratio.

$$\frac{V_0}{V} = \frac{R}{R + j(\omega L - 1/\omega C)}$$

$$\left|\frac{V_0}{V}\right| = \frac{R}{\{R^2 + (\omega L - 1/\omega C)^2\}^{1/2}}$$

$$\text{angle of } \frac{V_0}{V} = -\tan^{-1} \frac{\omega L - 1/\omega C}{R}$$

The frequency response curves for the RL and RC circuits were drawn with a frequency scale calibrated in multiples of ω_0. In this way they became *universal* curves which applied to any circuit of the appropriate type, no matter what the particular values of R and C (or L). It would also be useful to plot universal curves for the RLC circuit. To do this we can plot the curves in a scale of multiples of its centre frequency ω_0, but this is not enough. The shape of the curves will depend upon the ratio of the band width to ω_0, since ω_0 determines the scale.

$$\frac{W}{\omega_0} = \frac{R/L}{1/(LC)^{1/2}}$$

$$= R\left(\frac{C}{L}\right)^{1/2}$$

This ratio is simply in the reciprocal of the constant Q we found before.

* Note that this formula gives the band width in radians. In Hz the band width is $(1/2\pi)R/L$.

$$\frac{W}{\omega_0} = \frac{1}{Q}$$

or

$$Q = \frac{\omega_0}{W}$$

The Q of a resonant circuit is the ratio of its centre frequency to its band width.*

The effect of Q on the circuit shows very clearly in the frequency response curves. By juggling the magnitude and angle expressions somewhat to put them in terms of ω_0 and Q

$$\left|\frac{V_0}{V}\right| = \frac{1}{\{1 + Q^2(\omega/\omega_0 - \omega_0/\omega)^2\}^{1/2}}$$

$$\text{angle of } \frac{V_0}{V} = -\tan^{-1} Q\left(\frac{\omega}{\omega_0} - \frac{\omega_0}{\omega}\right)$$

These expressions are plotted in figure 33 for three values of Q. With these curves we can summarize the properties of Q.

The higher the Q of a resonant circuit:

(a) the sharper is its frequency response curve in the vicinity of ω_0;
(b) the narrower is its band width (for a given centre frequency);
(c) the larger is the voltage across the capacitor and inductor at resonance.

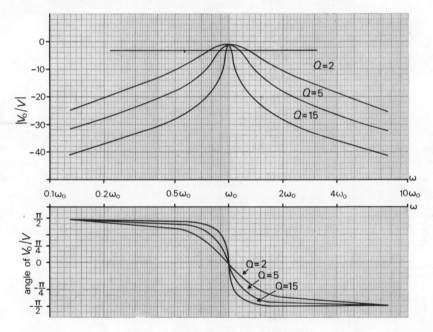

Figure 33 Frequency response curves for the series RLC circuit.

I have now shown two interpretations of the quality factor Q, one in terms of the ratio of the capacitor or inductor impedances to the total impedance at resonance, and one in terms of the bandwidth and the centre frequency. It can also be defined as the ratio of the maximum energy stored to the energy dissipated in one cycle. For the series RLC circuit these three interpretations yield the same expression, in terms of R, L and C. The third interpretation is the most general and can be used to describe the sharpness of resonances in systems other than electronic ones. It is also used to characterize other resonant circuits and as a 'quality' factor for capacitors and inductors since, in contrast to the idealized models we have discussed in this unit, real capacitors and inductors do dissipate some energy as well as store it.

* While this statement is correct for the Q of an RLC circuit as I have defined it, there are other definitions of Q and other resonant circuits for which this statement is only true if Q is large.

At what rate does the magnitude of the voltage ratio decrease as the frequency increases or decreases well away from the resonant frequency?

From figure 33, the magnitude of the voltage ratio decreases by 6 dB per octave both above and below ω_0 for frequencies well outside the pass band for *all* the values of Q. The slope of -6 dB per octave is the same as for the RL and RC circuits, which is as we should expect because it is the capacitor which causes the low frequency stop band and the inductor which causes the high frequency stop band.

SAQ4

Suppose you have a 1 mH inductor which has losses which can be represented by a series resistance of 10 Ω. Use this inductor to build a band pass circuit, as in figure 34, with a centre frequency of 10 kHz and a bandwidth of 3 kHz. (Note that these frequencies are *not* in radians.) How will the presence of the losses affect the maximum voltage ratio? Can you modify the circuit so that the bandwidth is 1 kHz?

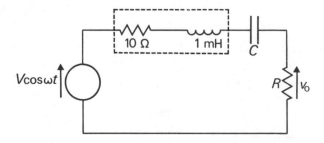

Figure 34 A band pass circuit using an inductor with losses represented by a series resistance.

The RLC circuit of figure 35 is a *parallel resonant circuit*. In contrast to the series resonant circuit discussed up to now it contains a current source in parallel with the inductor and capacitor. By the principle of duality its *current* ratio I_0/I has the same form as the voltage ratio of the series resonant circuit. It is a band pass circuit for currents.

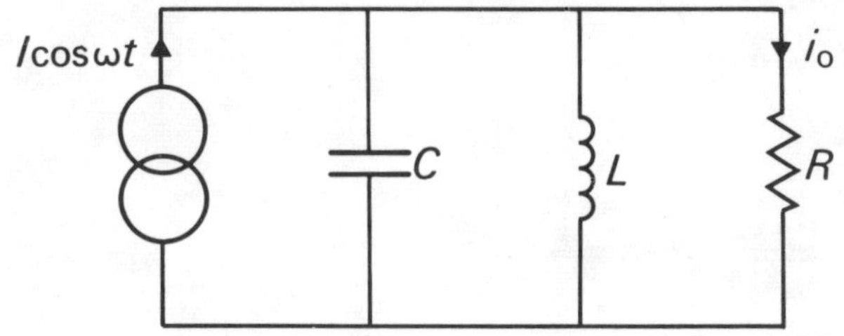

Figure 35 A parallel RLC circuit.

Section 6

7.6 Fourier series

Earlier in this unit, when I first introduced sinusoids, I asserted that they could be used as 'building block' functions. That is, by adding appropriate combinations of sinusoids, any desired nonsinusoidal signal could be formed. The importance of this property is that by adding the responses of a network to each of the sinusoidal components of a signal, we obtain the response to that signal itself (by the principle of superposition). In this way the response of a network to sinusoids of every frequency is all we need to know to find its response to other signals. In this section we will examine this assertion more closely.

Although according to mathematical theory the sinusoidal components of virtually any signal can be evaluated, in practice in very many applications it is not necessary to do so explicitly. This can best be explained by an example.

A high fidelity amplifier is designed to accept a signal from a source, such as a gramophone pick-up, with a very small maximum voltage and current. The amplifier produces a signal which is approximately a replica of the input signal multiplied by a constant called the gain. The gain must be large enough so that the output signal can drive a loudspeaker. To do this, the response of the amplifier to any of the sinusoidal components of the input signal must be to multiply the amplitude by the gain and leave the phase unchanged. It is impossible to build an amplifier which behaves this way at *all* frequencies. It is also impossible to find the sinusoidal components of the signals from all gramophone records which could be played through the amplifier. But it is possible to say that the frequencies of all these components must lie in a certain range. This range is about 30–20 000 Hz, which more than covers the range over which the human ear can hear. It is also possible to design an amplifier to have the desired response to all sinusoids with frequencies in this range.

The conclusion of this discussion is that the *idea* of sinusoidal components is necessary for the amplifier design because the design is based upon the amplifier's response to sinusoids even though it will normally be operated only with nonsinusoidal signals. The design doesn't require knowledge of the exact components of any one signal, just their frequency range.

The general subject of the analysis of signals into their sinusoidal components is more mathematical than is needed for this course, so I will only discuss a few representative signals to give you a general idea of what is involved.

First consider the triangular wave in figure 36(a). It is a periodic waveform which repeats every 2π seconds. Figures 36(b) & (c) overleaf show how this waveform can be built up by adding sinusoids. I have illustrated this for only half a cycle of the triangular wave. The other half cycle is symmetric to the first.

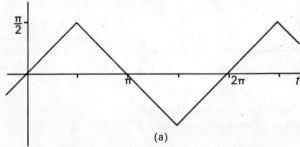

Figure 36 (a) *A triangular wave.*

The sinusoids in the left-hand side of the figure are components of the triangular wave. In the right-hand side each component is added to all those which preceded it to give successively better approximations to the triangular wave.

First let us look at the components needed to give these approximations. The first is a sinusoid with the same period as the triangular wave. (Only half of a cycle is shown.) The frequency at which any periodic waveform, such as the triangular wave, repeats is called its *fundamental frequency*. The sinusoidal component at that frequency is called the *first harmonic*.

fundamental frequency
first harmonic

The second component is a sinusoid at exactly three times the fundamental frequency. It is the *third harmonic*. (There is no second harmonic for this particular waveform.) Notice that its amplitude is much smaller than the amplitude of the first harmonic. Since the first harmonic alone is a fairly good approximation it is reasonable that the next component added to improve the approximation should be smaller.

The third component added is the *fifth harmonic*, a sinusoid at five times the fundamental frequency. Its amplitude is smaller still than the amplitude of the third harmonic.

The approximation formed by adding the first three terms is good enough so that the improvement made by adding the fourth term is not easy to see in the figure. The better the approximation is, the smaller are the successive terms needed to improve it. The fourth term is the *seventh harmonic* and is the smallest yet.

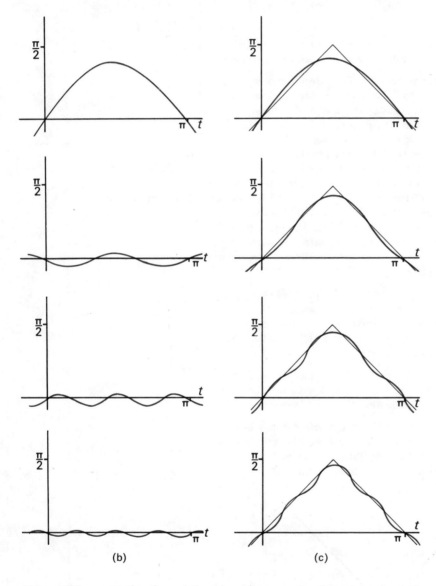

Figure 36 (b) the first four sinusoidal components of the triangular wave; (c) the successive approximations to the triangular wave formed by successively adding the components.

Although the approximation of the triangular wave by these four sinusoids appears quite good it is still not exact. It is clear that more terms need to be added to get an exact representation.

Now let us see what general principles are illustrated by this example.

1. The sinusoidal components of a periodic waveform are all at multiples of the fundamental frequency. That is, they are all harmonics.

2. The amplitude of successive terms gets progressively smaller*.

* For certain waveforms the amplitude of the first few terms may not get successively smaller, but the amplitude of later terms must.

3   Adding successive components improves the approximation but does not give an exact representation. In fact an infinite number of terms is generally required if an exact representation is needed.

If we let y represent the triangular waveform then the formula for y in terms of its sinusoidal components is

$$y = \frac{4}{\pi}\left(\sin t - \frac{1}{9}\sin 3t + \frac{1}{25}\sin 5t - \frac{1}{49}\sin 7t + \cdots\right) \qquad (4)$$

The ellipsis ($\cdots$) means that the series continues on in a similar manner. The method used to find the terms in equation (4) is given in the appendix. There are an infinite number of terms. Note that the amplitude of each is inversely proportional to the square of the frequency and so gets smaller on each successive term.

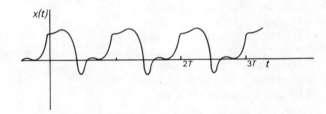

Figure 37 A periodic signal.

An expression such as equation (4) is called a *Fourier series* after its inventor, Joseph Fourier, a nineteenth century French scientist. The process of finding the Fourier series for a given periodic waveform is called *Fourier analysis*. Suppose we have a signal which repeats periodically as in figure 37. It repeats with a period of T seconds. Thus its fundamental frequency f_0 is

Fourier series

Fourier analysis

$$f_0 = \frac{1}{T}$$

In radians, the fundamental angular frequency ω_0 is

$$\omega_0 = 2\pi f_0$$
$$= \frac{2\pi}{T}$$

The basic postulate of Fourier analysis is that any periodic signal can be represented as a sum of sinusoids according to the formula*

$$x(t) = \frac{k_0}{2} + k_1 \cos(\omega_0 t + \phi_1) + k_2 \cos(2\omega_0 t + \phi_2)$$
$$+ k_3 \cos(3\omega_0 t + \phi_3) + \cdots + k_n \cos(n\omega_0 t + \phi_n) + \cdots \qquad (5)$$

A convenient notation for the Fourier series is

$$x(t) = \frac{k_0}{2} + \sum_{n=1}^{\infty} k_n \cos(n\omega_0 t + \phi_n)$$

where the sign $\sum$ means 'the sum of ...' and the notation $n = 1$ (below) and ∞ (infinity) (above) means that the expression after the summation sign is repeated once for every value of n in the range from one to infinity.

The first term is a constant $k_0/2$. (It can be thought of as a sinusoid of frequency zero times the fundamental frequency to give the formula consistency.) The constant represents the average or dc level of the signal. (Since the terms with nonzero frequency all have a dc level of zero they cannot contribute to the dc level of the signal.)

The constants k_0, k_1, k_2, etc., and ϕ_1, ϕ_2, ϕ_3, etc., are different for each signal. Finding the components of a signal really means finding these constants.

---

* *There are some restrictions on the type of signal to which this formula applies, but they don't often apply to signals used in electronics. See Ferris, C.D. (1962) Linear Network Theory, Columbus, Ohio; Merrill, pp. 137–8 for details.*

Find the constants k_0, k_1, k_2, etc., and ϕ_1, ϕ_2, ϕ_3, etc., for the triangular waveform. That is, put equation (4) into the same form as equation (5).

The constant term k_0 is zero. The other amplitudes are

$k_n = 0$ if n is even

$k_n = \dfrac{4}{\pi n^2}$ if n is odd

The phases of the terms whose amplitudes are not zero are alternately $-\pi/2$ and $\pi/2$. That is

$\phi_1 = -\pi/2 \quad \phi_3 = \pi/2 \quad \phi_5 = -\pi/2 \cdots$

In the beginning of this section I described a situation in which the actual Fourier components of a signal were not needed, merely their frequency range. However, from the present discussion you can see that since there are an infinite number of terms in a Fourier series, there is no highest frequency. Thus the range of frequencies is infinite. This difficulty can be overcome because the error caused by using only a finite number of terms as an approximation can be made as small as is desired simply by taking enough terms. Thus a practical 'highest frequency' can be specified once the maximum acceptable error for a given application is established.

If we want to find the frequency components of a signal which does not repeat itself periodically we cannot use a Fourier series. If you think of a signal which is not periodic as if it 'repeated' itself with an infinite period then its fundamental frequency (the reciprocal of the period) would be infinitesimally small. Multiples of this fundamental frequency would then include all frequencies, rather than just the discrete set ω_0, $2\omega_0$, etc., as for a periodic function. The sum in the Fourier series thus becomes an integral for nonperiodic functions

$$x(t) = \frac{1}{\pi} \int_0^\infty k(\omega) \cos\{\omega t + \phi(\omega)\} \, d\omega$$

This equation is called the *Fourier integral* for $x(t)$. Instead of finding the constants k_1, k_2, k_3, etc., and ϕ_1, ϕ_2, ϕ_3, etc., we must find the functions $k(\omega)$ and $\phi(\omega)$ to characterize the components of $x(t)$. Together $k(\omega)$ and $\phi(\omega)$ are the magnitude and phase of a complex function

Fourier integral

$$F(\omega) = k(\omega) \angle \phi(\omega)$$

called the *Fourier transform* of $x(t)$.

Fourier transform

Now let us look at another example, the Fourier series of the square wave shown in figure 38(a).

Its Fourier series is

$$v(t) = \sum_{n=1,3,5}^{\infty} \frac{4V}{n\pi} \cos\left(n\omega_0 t - \frac{\pi}{2}\right) \quad (6)$$

Or, in terms of sine functions

$$v(t) = \frac{4V}{\pi}\left(\sin \omega_0 t + \frac{1}{3}\sin 3\omega_0 t + \frac{1}{5}\sin 5\omega_0 t + \cdots\right)$$

If you are interested in how the coefficients were found, the steps are explained in Appendix 1.

Figure 38(b) shows the first four Fourier components of the square wave and the approximations formed by adding them successively. Notice that the amplitudes of the components decrease at a slower rate than for the triangular wave. From equation (6) they are inversely proportional to the frequency, rather than its square as for the triangular wave. This slower decrease is reflected in the fact that the approximations formed by corresponding numbers of terms are not as good for the square wave as for the triangular wave. The reason for this is the jump in the square wave. In general the magnitudes of the frequency components of a waveform decrease at a more rapid rate the 'smoother' the waveform is.

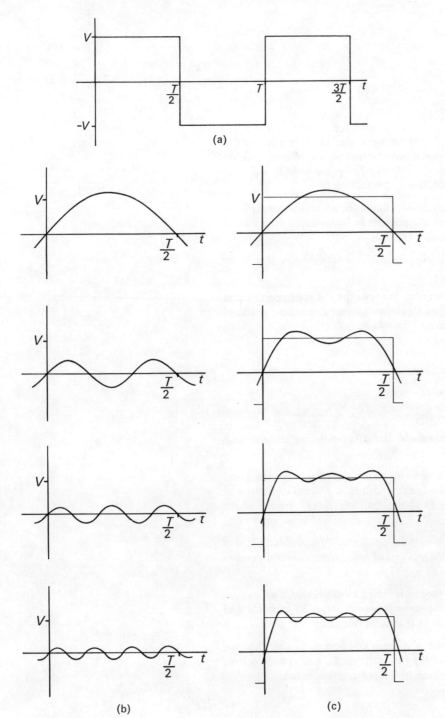

Figure 38 (a) A square wave; (b) the first four Fourier components of the square wave; (c) the successive approximations formed by successively adding the components.

Section 7

7.7 Summary of the unit

This unit described the response of networks composed of linear components to sinusoidal voltages and currents. It developed the techniques, phasor notation and complex impedance, which enable the sinusoidal response to be found without the need for solving the circuit's differential equation.

Section 7.2 describes the type of circuit, linear circuit, for which these techniques apply. Any circuit composed of linear components or linear approximations to nonlinear components is a linear circuit.

Any component with a characteristic curve which is a straight line is a linear component.

Currents and voltages in a linear circuit obey the principles of superposition and homogeneity. If x_1 and x_2 are two input signals to a linear circuit and y_1 and y_2 are corresponding output signals then:

Homogeneity

ay_1 is the response to ax_1

Superposition

$y_1 + y_2$ is the response to $x_1 + x_2$

The principle of superposition is especially useful in analysing circuits with several sources.

The 'building block' theory is based upon the principles of superposition and homogeneity. With it the response of a linear circuit to any signal can be found by combining the known responses to those 'building block' signals which combine to form that signal.

Section 7.3 describes the set of most used building blocks – the sinusoids. It also describes the response of linear components and of combinations of components to them.

The major reason why sinusoids are used as building blocks is that if all sources in a linear circuit are sinusoids of the same frequency, then all currents and voltages in the circuit will be sinusoids at that same frequency.

Therefore, to characterize the response of a linear component to sinusoids, we compare the amplitudes and phases of the current through, and voltage across, that component. The ratios of the voltage to current amplitudes are

    Resistor: R

    Capacitor: $1/\omega C$

    Inductor: ωL

The relative phases of the voltage and current are

    Resistor: in phase

    Capacitor: current leads by $\pi/2$

    Inductor: voltage leads by $\pi/2$

To find the response of a linear circuit to a sinusoidal source without using the differential equations of the circuit, we use phasors and complex impedances described in section 7.4. With them sinusoidal responses can be found using the same rules as for resistive networks.

A phasor is a complex number representing the phase and amplitude of a sinusoid. The complex impedance of a linear component is the ratio of the phasor of its voltage to the phasor of its current. For the three basic components it is

    Resistor: R Capacitor: $1/j\omega C$ Inductor: $j\omega L$

The impedances of capacitors and inductors vary with frequency in different ways while the impedance of a resistor is independent of frequency. Thus by combining resistors, capacitors and inductors we can form circuits which respond differently at different frequencies.

In section 7.5 we used the techniques developed in the earlier sections to find the expressions for the impedances and some voltage ratios in RC, RL and RLC circuits. We then examined the Argand diagrams and frequency response curves describing those impedances and ratios.

The RC and RL circuits could each be used as either high pass or low pass circuits. In their pass bands, the voltage ratios of all four circuits had magnitudes near 1 and angles near zero. In their stop bands the magnitude of the voltage ratio decreased by 6 dB per octave and the angle approached $+\pi/2$ (for a high pass circuit) or $-\pi/2$ (for a low pass circuit) as the frequency varied away from the pass band.

The cut off frequency is, by convention, the frequency at which the real and imaginary parts of the impedance are equal

RC circuit: $\omega_0 = 1/RC$
RL circuit: $\omega_0 = R/L$

By combining the behaviour of a low pass RL circuit and a high pass RC circuit, the RLC circuit is a band pass or resonant circuit. Its centre frequency is

$$\omega_0 = \frac{1}{(LC)^{1/2}}$$

The sharpness of the resonance depends upon the ratio of the impedances of the capacitor or inductor to that of the resistor near the resonant frequency. It is characterized by

$$Q = \frac{1}{R}\left(\frac{L}{C}\right)^{1/2}$$

Section 7.6 gave a brief introduction to the way in which sinusoids are used as 'building block' functions to represent signals. A periodic signal can be represented as a Fourier series, which is a sum of an infinite number of sinusoids whose frequencies are harmonics of the fundamental frequency of the signal. The amplitudes of these components get smaller as their frequencies get larger so that the signal can be approximated by a finite number of components.

Self-assessment questions

Question 5

Determine the voltage v in the circuit of figure 39. (Hint: The solution requires no mathematics!)

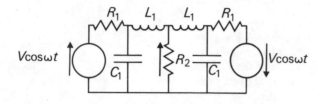

Figure 39

Question 6

Determine the voltage v in the circuit of figure 40.

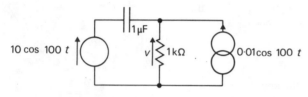

Figure 40

Question 7

Suppose a linear network contains only one source and that the source has a periodic waveform which is not sinusoidal. Give an argument to support the statement: 'All voltage and current waveforms in the circuit will be periodic with the same fundamental frequency as the source'.

Self-assessment answers and comments

Question 1

The third input is found from the first two as

$$v_{in(3)} = v_{in(1)} - v_{in(2)}$$

By the principle of homogeneity, the response to an input signal, $-v_{in(2)}$ is $-v_{out(2)}$. (See figure 41(a).)

By the principle of superposition

$$v_{out(3)} = v_{out(3)} + (-v_{out(2)})$$

which is shown in figure 41(b).

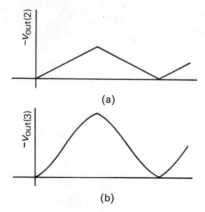

Figure 41

Question 2

Before answering the questions in detail, remember that the impedance of a capacitor is high at low frequencies and low at high frequencies. Thus at low frequencies the source voltage will appear mainly across the capacitor and at high frequencies mainly across the resistor.

1 The impedance of the RC circuit is

$$Z = R + \frac{1}{j\omega C}$$

$$= R - \frac{j}{\omega C}$$

An Argand diagram showing the impedance for three frequency values is in figure 42. As for the RL circuit, we denote by ω_0 that frequency at which the magnitudes of the capacitor's impedance and the resistor's impedance are equal

$$\left|\frac{1}{j\omega_0 C}\right| = |R|$$

$$\omega_0 = \frac{1}{RC}$$

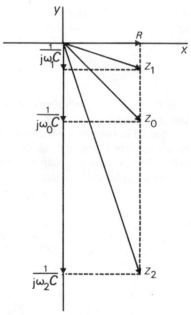

Figure 42

From the frequency response curves we shall see that ω_0 again represents the 3 dB frequency. At ω_2, a frequency lower than ω_0, we can see from figure 42 that the impedance is much nearer to that of the capacitor. At ω_1, a higher frequency than ω_0, the impedance is much nearer to that of the resistor.

2 Therefore most of the source voltage will appear as the output voltage for the circuit in figure 26(a) for frequencies above ω_0, and for the circuit in figure 26(b) for frequencies below ω_0. The circuit in figure 26(a) is a high pass circuit. That in figure 26(b) is a low pass circuit.

3 For figure 26(a)

$$\frac{V_0}{V} = \frac{R}{R + 1/j\omega C}$$

$$= \frac{j\omega CR}{1 + j\omega CR}$$

$$= \frac{j\omega/\omega_0}{1 + j\omega/\omega_0}$$

$$\left|\frac{V_0}{V}\right| = \frac{\omega/\omega_0}{\{1 + (\omega/\omega_0)^2\}^{1/2}}$$

angle of $\dfrac{V_0}{V} = \dfrac{\pi}{2} - \tan^{-1}\dfrac{\omega}{\omega_0}$

These are *precisely* the same expressions as for the high pass RL circuit. Therefore the magnitude and angle curves are those shown in figure 25.

For figure 26(b)

$$\frac{V_0}{V} = \frac{1/j\omega C}{R + 1/j\omega C}$$

$$= \frac{1}{1 + j\omega RC}$$

$$= \frac{1}{1 + j\omega/\omega_0}$$

$$\left|\frac{V_0}{V}\right| = \frac{1}{\{1 + (\omega/\omega_0)^2\}^{1/2}}$$

angle of $\dfrac{V_0}{V} = -\tan^{-1}\dfrac{\omega}{\omega_0}$

If you write the expressions for the frequency response of the low pass RL circuit in terms of its 3 dB frequency they will again be identical to these expressions for the RC circuit. Thus the curves in figure 23 also describe this circuit.

Question 3

1 The phasor for the voltage source is $10 + j0$. By the voltage divider rule the phasor V_0 for the voltage across the load resistor is

$$V_0 = \frac{1000}{1000 + 1000 + 1/j\omega C} \times 10$$

$$= \frac{10^4}{2 \times 10^3 + 1/j\omega C}$$

2 The circuit has a frequency response similar to that of the high pass RC circuit. At very high frequencies the impedance of the capacitor approaches zero (a short circuit) so V_0 is determined just from the resistors and is a maximum. By the voltage divider rule

$$V_{0(\text{max})} = \frac{1000}{1000 + 1000} \times 10$$

$$= 5$$

(The phasor is 5. The sinusoidal voltage is $5 \cos \omega t$.)

3 The ratio of V_0 to its maximum value is

$$\frac{V_0}{V_{0(\text{max})}} = \frac{2 \times 10^3}{2 \times 10^3 + 1/j\omega C}$$

You should recognize this expression from the RC high pass circuit. The 3 dB frequency is the frequency at which the real and imaginary terms in the denominator have the same magnitudes.

$$2 \times 10^3 = \frac{1}{\omega_0 C}$$

or

$$\omega_0 = \frac{1}{2 \times 10^3 C}$$

The 3 dB frequency must be lower than 1 kHz (or $2\pi \times 10^3$ rad s^{-1}), i.e.

$$2\pi \times 10^3 > \frac{1}{2 \times 10^3 C}$$

Thus

$$C > \frac{1}{4\pi} \times 10^{-6}\ \text{F}$$

or

$$C > \frac{1}{4\pi}\ \mu\text{F}$$

Question 4

The centre frequency determines the value of the capacitor needed.

$$C = \frac{1}{\omega_0^2 L}$$

$$= \frac{1}{(2\pi \times 10^4)^2 \times 10^{-3}}$$

$$= 0.25\ \text{pF}$$

The band width determines the resistance needed.

$$W = \omega_0/Q$$

$$= R/L$$

so

$$R = WL$$

$$= 2\pi \times 3000 \times 10^{-3}$$

$$= 18.7\ \Omega$$

The inductor resistance accounts for $10\ \Omega$. Thus we add an additional $8.7\ \Omega$. The circuit appears as in figure 43.

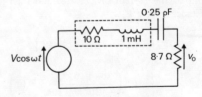

Figure 43

At resonance the impedances of the capacitor and inductor cancel each other, but the series resistance of the inductor remains. The maximum voltage ratio is found from the voltage divider rule to be

$$\left|\frac{V_0}{V}\right|_{(\text{max})} = \frac{8.7}{8.7 + 10}$$

$$= 0.47$$

Without the inductor resistance the voltage ratio would be one, as we saw previously.

For a band width of 1 kHz, a resistance of 6.3 Ω is needed. Since the inductor resistance is greater than that, a filter with a 1 kHz band width cannot be built using the inductor described.

Question 5

The voltage v can be found very easily using the principle of superposition. First consider the voltage at v due to the left-hand voltage source alone. Call this voltage v_1. Next consider the voltage due to the right-hand voltage source alone. Call it v_2. Notice that the circuit is symmetric about the central resistor. Thus the two circuits formed by setting either one of the two sources to zero are electrically the same except that the reference directions of the sources are reversed. Thus by the principle of homogeneity $v_1 = -v_2$. The total voltage v is thus

$$v = v_1 + v_2$$
$$= 0$$

Question 6

The catch to this problem is that the two sources are at different frequencies. We can use superposition to find the phasors of the response of the two sources separately, but we cannot then add the two phasors to form a single phasor.

The phasor for the voltage source is $10 + j0$. If V_1 represents the phasor of the voltage appearing at v due to it, then

$$V_1 = \frac{R}{R + 1/j\omega C} \times 10$$

$$= \frac{10^3}{10^3 - j/10^{-3}} \times 10$$

$$= \frac{10}{1 - j}$$

The corresponding sinusoidal voltage is

$$v_1 = \frac{10}{\sqrt{2}} \cos\left(1000t + \frac{\pi}{4}\right)$$

The phasor for the current source is $0.01 + j0$. Notice that there is no way of telling from their phasors that the frequencies of the two sources are different. Let V_2 be the phasor of the voltage at v due to the current source. The current through the resistor is found by the current divider rule and then multiplied by the resistance to give V_2.

$$V_2 = R \left(\frac{1/j\omega C}{1/j\omega C + R} \times 0.01\right)$$

$$= 10^3 \left(\frac{-j/10^{-4}}{-j/10^{-4} + 10^3} \times 0.01\right)$$

$$= 10 \left(\frac{-j}{-j + 0.1}\right)$$

Its magnitude and angle are

$$|V_2| = 10 \times \frac{|-j|}{|-j + 0.1|}$$

$$= \frac{10}{1.005}$$

$$= 9.95$$

angle of $V_1 = -\frac{\pi}{2} - \tan^{-1}\frac{-1}{0.1}$

$$= -\frac{\pi}{2} + 1.47$$

$$= -0.12 \text{ rad}$$

The corresponding sinusoidal voltage is

$$v_2 = 9.95 \cos(100t - 0.12)$$

Thus the total voltage v is

$$v = v_1 + v_2$$
$$= \frac{10}{\sqrt{2}} \cos\left(1000t + \frac{\pi}{4}\right) + 9.95 \cos(100t - 0.12)$$

Question 7

Since the source waveform is periodic it can be represented by a Fourier series, i.e. a sum of sinusoids whose frequencies are harmonics of the fundamental frequency.

By the principle of superposition, the response of any voltage or current in the circuit to the source waveform is the sum of the responses to each frequency component. The response to each frequency component is a sinusoid at the frequency of that component. (This is the key point, because the only waveform for which we know the response in general terms is a sinusoid.) Thus the response also consists of a sum of sinusoids which are harmonics of the fundamental frequency.

Since a harmonic has a frequency which is a multiple of the fundamental frequency it also repeats *at* the fundamental frequency. (For example, the third harmonic repeats at the fundamental frequency, three cycles at a time.)

Thus the total response of any voltage or current in the circuit to any periodic waveform is the sum of sinusoids which repeat at the fundamental frequency. It is thus periodic with the same fundamental frequency.

Appendix 1 Obtaining Fourier coefficients

A signal $x(t)$ can be represented as a Fourier series

$$x(t) = \frac{k_0}{2} + \sum_{n=1}^{\infty} k_n \cos(n\omega_0 t + \phi_n)$$

where k_n is the amplitude and ϕ_n the phase of each harmonic component. By using the identity

$$k \cos(\omega t + \phi) = a \cos \omega t + b \sin \omega t$$

where $a = k \cos \phi$, $b = -k \sin \phi$, or $k = (a^2 + b^2)^{1/2}$, $\phi = \tan^{-1}(-b/a)$, the Fourier series can also be written as

$$x(t) = \frac{k_0}{2} + \sum_{n=1}^{\infty} (a_n \cos n\omega_0 t + b_n \sin n\omega_0 t)$$

These two forms of a Fourier series correspond to the magnitude and angle, and real and imaginary forms of a phasor respectively.

The constants k_0, a_1, b_1, a_2, b_2, etc., can be found by using a very convenient property of sinusoids. Multiply by $\cos m\omega_0 t$ and integrate over one period.

$$\int_0^T x(t) \cos m\omega_0 t \, dt$$

$$= \int_0^T \left\{ \frac{k_0}{2} + \sum_{n=1}^{\infty} (a_n \cos n\omega_0 t + b_n \sin n\omega_0 t) \right\} \times \cos m\omega_0 t \, dt$$

The right-hand side of this equation is the integral of an infinite sum of terms. However, the integral of each term is zero except for the term containing $\cos n\omega_0 t$ with $n = m$. For that term the integral is $a_n T/2$. Thus

$$\int_0^T x(t) \cos n\omega_0 t \, dt = \frac{a_n T}{2}$$

or

$$a_n = \frac{2}{T} \int_0^T x(t) \cos n\omega_0 t \, dt$$

In a similar manner

$$b_n = \frac{2}{T} \int_0^T x(t) \sin n\omega_0 t \, dt$$

$$k_0 = \frac{2}{T} \int_0^T x(t) \, dt$$

The constants k_1, ϕ_1, k_2, ϕ_2, etc., are then found from

$$k_n = (a_n^2 + b_n^2)^{1/2}$$

$$\phi_n = \tan^{-1} \frac{-b_n}{a_n}$$

For the square wave

$$k_0 = \frac{2}{T} \int_0^T v(t) \, dt$$

$$= \frac{2}{T} \left\{ \int_0^{T/2} V \, dt + \int_{T/2}^T (-V) \, dt \right\}$$

$$= 0$$

$$a_n = \frac{2}{T} \int_0^T v(t) \cos n\omega_0 t \, dt$$

$$= \frac{2}{T} \left\{ \int_0^{T/2} V \cos n\omega_0 t \, dt + \int_{T/2}^T (-V) \cos n\omega_0 t \, dt \right\}$$

$$= 0$$

$$b_n = \frac{2}{T} \int_0^T v(t) \sin n\omega_0 t \, dt$$

$$= \frac{2}{T} \left\{ \int_0^{T/2} V \sin n\omega_0 t \, dt + \int_{T/2}^T (-V) \sin n\omega_0 t \, dt \right\}$$

$$= \frac{2}{T} \left(\frac{VT}{n} + \frac{VT}{n} \right) = \frac{4V}{n} \quad \text{if } n \text{ is odd}$$

or

$$= \frac{2}{T}(0 + 0) = 0 \quad \text{if } n \text{ is even}$$

Therefore

$$k_n = (a_n^2 + b_n^2)^{1/2}$$

$$= b_n$$

$$= \frac{4V}{n} \quad \text{if } n \text{ is odd}$$

$$= 0 \quad \text{if } n \text{ is even}$$

$$\phi_n = \tan^{-1} \frac{-b_n}{a_n}$$

$$= \tan^{-1} \frac{-b_n}{0}$$

$$= \frac{-\pi}{2}$$

For the triangular wave, $T = 2\pi$, $\omega_0 = 1$. The wave can be represented for one period as

$$y = t \qquad \text{for } 0 < t \leq \pi/2$$
$$= \pi - t \qquad \text{for } \pi/2 < t \leq 3\pi/2$$
$$= 2\pi + t \qquad \text{for } 3\pi/2 < t \leq 2\pi$$

Since the average value of the waveform is zero, $k_0 = 0$.

$$a_n = \frac{1}{\pi} \int_0^{\frac{\pi}{2}} t \cos nt \, dt + \frac{1}{\pi} \int_{\frac{\pi}{2}}^{\frac{3\pi}{2}} (\pi - t) \cos nt \, dt$$

$$+ \frac{1}{\pi} \int_{\frac{3\pi}{2}}^{2\pi} (2\pi + t) \cos nt \, dt$$

Evaluating the integrals gives $a_n = 0$.

$$b_n = \frac{1}{\pi} \int_0^{\pi/2} t \sin nt \, dt + \frac{1}{\pi} \int_{\frac{\pi}{2}}^{\frac{3\pi}{2}} (\pi - t) \sin nt \, dt$$

$$+ \frac{1}{\pi} \int_{\frac{3\pi}{2}}^{2\pi} (2\pi + t) \sin nt \, dt$$

Evaluating the integrals gives

$$b_n = \frac{4}{\pi n^2} \sin n\pi/2$$

but since

$$\sin \frac{n\pi}{2} = 1 \qquad \text{for } n = 1$$
$$= 0 \qquad \text{for } n = 2$$
$$= -1 \qquad \text{for } n = 3$$
$$= 0 \qquad \text{for } n = 4$$
$$= 1 \qquad \text{for } n = 5$$

etc.

$$b_n = 0 \qquad \text{for even } n$$
$$= 1 \qquad \text{for } n = 1, 5, 9, \text{etc.}$$
$$= -1 \qquad \text{for } n = 3, 7, 11, \text{etc.}$$

Electromagnetics and Electronics

1. Charges at rest and in motion
2. Magnetism and electromagnetic induction
3. Flux and magnetic circuits
4. Magnetic material and electrostatics
5. Semiconductor devices
6. Signal processing I. Transient response
7. Signal processing II. Sinusoidal response
8. Network modelling and analysis
9. Ac measurement and power
10. Power supplies
11. Amplifiers and amplification
12. } Design study I. Power amplifier
13. }
14. } Design study II. Time base generator
15. }
16. Design study III. Servomechanism
17. Design study IV. Characteristic curve tracer